THE RIGHT CHEMISTRY

THE RIGHT CHEMISTRY

108 Enlightening, Nutritious, Health-
Conscious and Occasionally Bizarre Inquiries
into the Science of Everyday Life

JOE SCHWARCZ, PhD

DOUBLEDAY CANADA

Doubleday Canada and colophon are registered trademarks

Library and Archives Canada Cataloguing in Publication

Schwarcz, Joe
The right chemistry / Joe Schwarcz.

Issued also in electronic format.
ISBN 978-0-385-67159-0

I. Chemistry—Popular works. 2. Chemistry—Miscellanea.
I. Title.

QD37.S38153 2012   540   C2012-902442-2

Cover image: Shawn Hempel | Dreamstime.com
Cover design: Andrew Roberts
Printed and bound in the USA

Published in Canada by Doubleday Canada,
a division of Random House of Canada Limited

Visit Random House of Canada Limited's website: www.randomhouse.ca

I0 9 8 7 6 5 4 3 2 I

CONTENTS

Introduction   *vii*

First, the Past   I

Tricks of the Trade   25

Creatures Great and Small   43

Claptrap   65

To Your Health   99

Bites and Sips   127

Mirror, Mirror on the Wall   173

Weird and Wonderful   191

Chemicals for Better and for Worse   211

Chemistry Up Close   251

How Green Was My Chemistry   261

Index   275

INTRODUCTION

I finally got my way. This book's title is one I first pleaded for almost fifteen years ago when I wrote my first book. I was bothered by the negative image that chemistry had acquired and I wanted to set the record straight. My message then, as it is now, was that there are no safe or dangerous chemicals, only safe or dangerous ways to use them. To me, it was all about making the right decisions, knowing the "right chemistry." And I also had in mind that people used this expression in a positive, albeit inappropriate fashion to describe the relationship between two people, or between players on a team.

To my dismay, I couldn't get publishers to bite. Including *chemistry* in the title is the kiss of death, they maintained. It was their view that people associate chemicals with toxins, and that for too many, chemistry brings back memories of a dreadful high school experience. So we came up with titles like *Radar, Hula Hoops and Playful Pigs; The Genie in the Bottle; That's the Way the Cookie Crumbles; Dr. Joe and What You Didn't Know; The Fly in the Ointment; Let Them Eat Flax; An Apple a Day; Brain Fuel; Science, Sense and Nonsense; Dr. Joe's Brain Sparks;* and *Dr. Joe's Health Lab.* In truth, all these books were about the right chemistry, which to me means using our knowledge of molecular behaviour to decipher chemical phenomena. And virtually everything in life comes down to chemical phenomena. The foods we eat, the medications we take, the cosmetics we apply, the energy we generate, the fabrics we wear, the plastics we use and the very workings of the human body surrender their mysteries to insightful chemical knowledge.

My intent is not to be an uncritical cheerleader for chemicals. After all, chemicals are nothing other than the inanimate building blocks of all matter. We are awash in some sixty million of them, the majority created by nature, a minority by chemists. They're not good or bad, not safe or dangerous. Their effect on us depends on how they are used, and what our exposure to them is. There is much debate about the impact of some chemicals on our lives, but about one issue there should be no debate: chemicals are interesting! So let's whet your appetite for learning about them with a few tantalizing morsels.

If you've ever battled moths or nasty household smells, you've probably met para-dichlorobenzene. A registered insecticide, it does away with moths very nicely. It can also do away with smells in urinals, either by masking the odour with its own strong smell or by preventing bacteria in stagnant urine from forming ammonia and other malodorous amines. That's why it is a common ingredient in those urinal cakes that make for such inviting tinkle targets. You'll also find para-dichlorobenzene in various air fresheners. And since we breathe the air, you'll find it in your blood. And what is it doing there? Probably not much good.

Para-dichlorobenzene is a known animal carcinogen and a suspected human one as well. Based on some elegant work using a tiny worm known as *C. elegans,* the suspicion is that this compound, as well as its chemical cousin naphthalene, can cause cancer by blocking apoptosis, a programmed process of "cell suicide" that occurs in all living organisms. Apoptosis is a sort of brake to prevent unchecked cellular proliferation, as happens in cancer. I wouldn't worry about peeing on a urinal cake, but I'd rather not make a habit of sniffing para-dichlorobenzene elsewhere.

Now on to piperine, the compound responsible for the flavour of black pepper, the most widely used spice in the world. First cultivated in India some four thousand years ago, pepper became a much sought-after commodity because of its ability to counter the offensive odours and tastes that were common before effective food preservation techniques were introduced. In the sixteenth century, pepper was so valuable that dock workers in England were prohibited from wearing clothes with pockets or cuffs for fear that they would abscond with a few corns.

America's first millionaire, Elias Derby of Salem, made his fortune by importing peppercorns and then used the money to endow Yale University. And now it is from Yale that we get a report of piperine's potential role in breast cancer prevention. A combination of piperine and curcumin, a compound isolated from turmeric, when applied to cultured breast cancer cells in the lab, has the effect of decreasing the number of stem cells. Limiting stem cells limits the number of cells that have a potential to form tumours. Of course, such laboratory research rarely translates into effective practice when it comes to people, but the concept that cancer risk can be reduced with dietary compounds that have very low toxicity is an appealing one.

Acetone peroxide is, I hope, a chemical you won't encounter. It's an explosive, and a powerful one at that. It was used in the infamous London bombings in 1907 and was also the substance that Richard Reid, the "shoe bomber," was trying to ignite to trigger an explosion that could have brought down an American Airlines flight in

2001. Luckily, Reid had been trudging around in rainy weather and the fuse in his shoe failed to ignite. He won't be getting another chance to experiment with explosives, seeing as he'll be spending the rest of his life in jail. But you can thank Reid for having to remove your shoes when you go through airport security.

Unfortunately, acetone peroxide is not difficult to make, requiring only acetone, hydrogen peroxide, a strong acid and a modicum of chemical knowledge. A highly unstable compound, it readily decomposes with heat or impact into acetone vapour and ozone. Hundreds of litres of gas can be produced from a few hundred grams of solid material in a fraction of a second! The shock wave created by this rapid release of gases is what we term an *explosion.* Unlike most explosives, acetone peroxide contains no nitrogen, making it "transparent" to airport detection equipment that tests for nitrogen-based residues. No wonder that acetone peroxide has earned the nickname "Mother of Satan."

Androstenone is not hard to find. All you have to do is raise your arms. It's right there in underarm sweat. And depending on your genetic makeup, you may or may not be able to sniff it. Roughly 25 per cent of the population cannot smell androstenone at all, 40 per cent think it smells like rotten wood or urine, while the rest find it to be a pleasing fragrance. Female pigs, on the other hand, are all sensitive to androstenone. Actually, more than just sensitive; if they are in heat and they smell androstenone, they will immediately assume a mating stance. That's because androstenone is a pig pheromone—in other words, a chemical that triggers a social response in another member of the same species.

The male pig produces androstenone and wafts it into the air from its saliva whenever it is in the "mood"—which, as for most males, is always. Boarmate is a commercially available androstenone spray that farmers can use to get their female pigs into the right frame of mind before disappointing them with an artificial-insemination rod. Androstenone's presence in truffles explains why female pigs can be used to hunt for these underground delicacies. Of course, only the 35 per cent or so of consumers who like the smell of androstenone will enjoy truffles—the rest are wasting their money on the $1,600-a-pound fungi.

Androstenone's effect on the sex lives of pigs and its occurrence in human sweat has given rise to conjecture that it may be a human pheromone as well. A number of companies have jumped the gun and now sell various androstenone preparations claiming that they help attract women. They claim androstenone helps foster "the right chemistry"! Maybe for pigs.

Here, I'd like to concentrate on the right chemistry for people. And the powers that be have agreed to the title I've long been craving, hopefully because, over the last few decades, I've had some success in making the point that the term *chemical* is not synonymous with *toxin* and that chemistry is not to be feared, it is to be understood. It is with a view towards providing such an understanding that I've crafted these questions and answers that I hope portray the amazingly broad scope of chemistry and emphasize the importance of getting that chemistry right.

FIRST,
THE PAST

A paper entitled "A new method of treating neuralgia by the direct application of opiates to the painful points" was published in 1855. What was this new method?

The use of a hypodermic syringe to inject a painkiller. In this landmark publication, published in the *Edinburgh Medical and Surgical Review*, Alexander Wood described the successful injection of morphine through a hollow needle attached to a syringe. Syringes of various types had been around since the time of the ancient Romans, but had never been used to inject a painkilling drug into the body.

Its name deriving from the Greek word for "tube," a syringe is a device consisting of a hollow cylinder with a tightly fitting plunger designed to take in and expel a liquid or a gas. Starting with the Romans, syringes were used to cleanse various bodily orifices. Enema syringes enjoyed special popularity, catering to the belief that illnesses stemmed from a dirty colon. Bladder-cleansing syringes, which were equipped with a long tube that had to be threaded up the urethra, mercifully made a late entry and quickly disappeared.

Early attempts at blood transfusion led to the attachment of a hollow-pointed needle to the syringe, which then triggered the idea of injecting drugs by means of a syringe.

The chance discovery that iron perchlorate had the ability to coagulate blood led to early attempts to prevent aneurysms from bursting by injecting the dilated blood vessels with this chemical. And then along came Dr. Wood with his idea of injecting morphine by means of a hypodermic syringe. The same year that Wood published his landmark paper, German chemist Friedrich Gaedeke discovered a compound in coca leaves that was subsequently isolated and named "cocaine" by Albert Niemann. A Peruvian army physician, Thomas Moreno y Maíz, experimented further with cocaine by injecting it into his leg. Why he did this isn't clear, but he did notice that injection of cocaine resulted in insensitivity. He even went on to suggest the use of such injections for local anaesthesia, which was brought to fruition by the Viennese ophthamologist Karl Koller. Cocaine became a widely used local anaesthetic.

Today, hypodermic injections are an essential part of medicine, their use ranging from vaccines to painkillers. But Wood's introduction of opiate injection for pain had an undesirable consequence as well. It made possible a new type of opiate abuse, intravenous injection of illegal drugs, that has haunted us since.

What are Prince Rupert's Drops and how are they connected to hockey?

We can take for granted that Prince Rupert of the Rhine, Duke of Bavaria, 1st Duke of Cumberland was not a hockey fan. He lived in the seventeenth century, some two hundred years before the first organized hockey game was played at Montreal's Victoria Skating Rink on March 3, 1875, between two teams made up of McGill University students. Still, Rupert's name is connected with the game. And it is also connected with my exploding glass patio table (more about which later). That's because Prince Rupert's Drops are glass beads that form when molten glass is dripped into cold water.

Actually there is a little controversy about that first hockey game. Both Kingston, Ontario, and Windsor, Nova Scotia, claim that a version of pond hockey was played there before the formal game in Montreal. It is also possible that indigenous people were the first to play a game similar to hockey even before Europeans settled in North America.

One thing, however, is for sure: wherever that first game was played, there was no protective glass around the rink. Now, of course, spectators can sit in the front row and have a clear view of the action without having to duck the puck.

Protective glass surrounding the boards first appeared in the NHL at Toronto's Maple Leaf Gardens in 1948 and became mandatory in all National Hockey League arenas in the 1950s. And this is where Prince Rupert comes into the picture. Or at least the glass drops named after him do. "Prince Rupert's drops" were the forerunners of the tempered glass panels that, until this year, protected spectators from pucks and errant sticks. Now those glass panels are being replaced with acrylic ones that are more flexible and have more "give," reducing the chance of injuries when players are smashed into them.

Prince Rupert's Drops feature a unique property. The bulbous end of the solidified tear-shaped drop readily withstands blows from a hammer, while just the slightest disturbance of the thin tail results in an explosive disintegration of the glass. Supposedly, the beads were the invention of Prince Rupert, son of German Prince Frederick V and Elizabeth, daughter of James I of England.

Rupert was an interesting man, to be sure. He had a varied career as a soldier, admiral, colonial governor and scientist. In fact, he was one of the founders of the Royal Society in Britain and even set up a laboratory in Windsor Castle.

It was during a meeting of the Royal Society that Rupert demonstrated the strange properties of his glass beads to his cousin King Charles II, stimulating the king to entertain members of his court by having subjects hold the bulb end in their palms, and then break off the tip. The startled victims watched the glass harmlessly explode into a powder. While Rupert did introduce his "drops" to England, it is unlikely that he invented them. Tempered glass was already known to European artisans, and Rupert had probably encountered it during his military trysts across the continent.

So, what is going on here? Glass is made by melting and then cooling sand, which essentially is silicon dioxide. The properties of the finished product are to a large extent determined by the rate of cooling. When molten glass is dropped into cold water, the rapid cooling of the outside of the drop and the slower cooling on the inside results in the formation of a network of linked silicon and oxygen atoms that provides great impact resistance. But it also produces internal stresses that can cause a sudden disintegration into tiny pieces, commonly precipitated by a small fracture that then spreads quickly through the whole glass.

We've seen hockey players and spectators showered with tiny fragments of glass after impact with a puck or body. And some of us have experienced the spontaneous explosion of other items made from tempered glass. Like patio tables.

Recently, my wife and I were roused by a giant bang coming from the backyard. We rushed outside to find the patio covered with myriad pieces of broken glass. The last time I was confronted by such a scene was after some lout broke into my car to steal my GPS. But in this case, there was no apparent criminal activity. This had been a spontaneous explosion.

The usual explanation offered is that changes in temperature cause slight expansion and contractions of the glass, adding to the built-in stresses. A catastrophic failure can then occur, triggered either by one final temperature change or even by a slight impact.

As far as hockey is concerned, the problem with tempered glass is not that it can crumble into a million pieces. The flying pieces are too small to cause an injury and the mess can be readily cleaned up. But the tempered glass is very hard, and being bodychecked into it can feel like hitting a rock wall. Worse than that, the lack of give can result in injuries to the body or head.

The appeal of using tempered glass around hockey rinks has been the clear visibility it provides. Panels can be installed without any seams, and the tempered glass is very resistant to scratches. Those advantages, however, have to be weighed against the greater safety provided by more flexible protection available in the form of acrylic panels.

Polymethyl methacrylate, commonly known as Plexiglas, has been around since the 1930s. It is a very tough plastic and offers excellent protection for spectators while providing far more cushioning for players than tempered glass. The downside is that it scratches easily, and the scratches interfere with visibility. That's why it has only been sporadically used in hockey arenas. But with an increased focus on head injuries, National Hockey League rinks are replacing tempered glass with acrylic panels. These panels can be laminated with a thin layer of polycarbonate, a plastic that resists scratches.

Now, if we could only find a way of stopping some of those hard-headed players from bashing opponents' soft faces into the acrylic panels.

They were made of silk and helped thousands of captured soldiers and downed pilots escape from German prisoner of war camps during the Second World War. What were they?

Maps. Printed on silk, the maps were remarkably clear and detailed, and above all, easy to conceal. The idea of printing on silk was not new—thousands of years ago, the ancient Chinese were already drawing maps of trade routes on silk. By the eighteenth century, printing on silk had become commonplace enough to make souvenir scarves with maps of spa towns a popular commercial item. But it was during WWII that printing on silk became a truly valuable process. British air crews carried silk maps in their survival kits to help them escape if shot down. Unlike paper, silk maps were waterproof and could be easily hidden in clothing to avoid detection. They are not damaged by repeated folding and unfolding. The problem was to achieve the detailed printing that was necessary. This is where MI9, the escape and evasion wing of British Military Intelligence, teamed up with the Bartholomew map company, and, amazingly, with John Waddington Ltd., a printer and board game manufacturer that also happened to be the U.K. licensee for the Parker Brothers board game Monopoly.

Before the war, Waddington had been printing theatre programs for special events on silk and had developed a great deal of experience with silk printing. It was the addition of pectin to the inks normally used that made clear, detailed printing possible. And Waddington was an ideal company for British Intelligence to work with because the Germans allowed board games to be sent to prisoners of war. The company managed to hide silk maps in the board of special editions of Monopoly along with a miniature compass. The enormously detailed maps showed cities, towns, villages, lakes, rivers, population densities, elevations, roads, railroads, mountain passes, military installations, and sometimes even ocean currents and navigational star charts. It is estimated that the maps helped about ten thousand prisoners of war escape. In 1942, the Americans got into the game and escape maps were standard issue to all U.S. Air Force personnel.

Even with all the modern technology available, silk maps still retain their importance. American pilots were issued silk maps during the Gulf War. Unfortunately, none of the special-edition Monopoly games that were delivered to prisoners of war have survived, but some of the silk maps they concealed are on view in museums.

In 1909, the U.S. government launched a lawsuit that would be referred to as The United States vs. Forty Barrels and Twenty Kegs of Coca-Cola. What did the government want the company to remove from its product?

The forty barrels and twenty kegs did not have to take the witness stand, but numerous experts were called to testify in the lawsuit that was aimed at forcing Coca-Cola to rid the beverage, not of cocaine, as one might think, but of caffeine. In 1906, the U.S. Congress passed the Pure Food and Drugs Act, which was the first legal attempt to curb the sale of the numerous patent medicines promoted with colourful but unsubstantiated claims.

Two men played a large part in prompting the passage of the law: Samuel Hopkins Adams, whom today we would call an investigative journalist, and Dr. Harvey Washington Wiley, who was chief of the Bureau of Chemistry, the forerunner of the FDA. Adams published a series of articles in *Collier's Weekly* under the title

"The Great American Fraud," alerting the public to the hazards of morphine, heroin, alcohol, cocaine and cannabis, all common ingredients in patent medicines, while Wiley was a passionate crusader against what he considered was adulteration of the food supply with the likes of bleached flour, the sweetener saccharin, the preservative sodium benzoate and the stimulant caffeine.

The Act forbade the sale of adulterated foods and "poisonous" patent medicines, although it was vague about what these terms encompassed. No specific ingredients were made illegal, but the Act required the listing of all ingredients on the label. The idea was that if people saw that the products contained substances that Harvey's Bureau of Chemistry had listed as possibly deleterious, they would avoid them. Indeed, the sales of patent medicines that contained opiates dropped by some 33 per cent after passage of the Pure Food and Drugs Act.

Wiley was convinced, without any evidence, that caffeine was deleterious to health and wanted it removed from the food supply. He took aim at Coca-Cola, a beverage that had sprouted its roots in the patent-medicine era and had become an immensely popular beverage. "Coke," as it came to be called, was the brainchild of Atlanta druggist John Pemberton, who was looking for a remedy for his customers' aches and pains. The properties of caffeine, found in the African kola bean, and cocaine, extracted from South American coca leaves, were already known at the time, and Pemberton combined them with alcohol and sugar. Pemberton had a soda fountain in his drugstore and came up with the brilliant idea of diluting his syrupy, caramel-coloured concoction with carbonated water in front of the customers and selling it as a remedy for aches and pains. To his surprise, it quickly developed a reputation as a hangover cure, and sales took off. Pemberton would never capitalize on the popularity of his invention. In failing health and addicted to morphine, he sold his formula for the paltry sum of $2,300 to Asa Candler, who would make Coca-Cola a corporate success and its logo the most recognizable trademark in the world.

Since Congress had classified cocaine as a deleterious substance, Candler removed it from the formula even though there was no legal obligation to do so. He still used coca leaves for flavouring, but only once their cocaine content was removed. Candler was a teetotaller and also voluntarily reduced the alcohol content of his beverage. But when it came to caffeine, he stood his ground against Wiley's rants. After all, caffeine had not been classified as "deleterious," and Candler considered it to be an integral part of Coca-Cola's appeal. Wiley, though, was adamant about caffeine being a dangerous drug, and since the Food and Drugs Act stated that dangerous drugs could not be sold, he set out to prove the compound's ability to endanger health. But the problem was that there was no scientific evidence to implicate caffeine. So Wiley manufactured some. He fed the media and government officials false reports of caffeine's ill effects and managed to trigger the lawsuit that became famous as *The United States vs. Forty Barrels and Twenty Kegs of Coca-Cola*, because the affair began with government agents seizing forty barrels and twenty kegs of Coca-Cola syrup in Chattanooga, Tennessee.

Wiley brought out the heavy artillery for the trial. He enlisted prominent evangelist George Stuart to testify that drinking Coca-Cola had resulted in "wild nocturnal freaks, violations of college rules and female proprieties, and even immoralities" at a girls' school. Coke's attorneys fought back, pointing out that coffee and tea had more caffeine than Coca-Cola. Wiley's passion got the better of him as he sought out expert witnesses who, for large sums, would testify about the dangers of caffeine. When the judge eventually found in favour of Coca-Cola, Wiley's reputation was sullied, leading to his resignation from the Bureau of Chemistry. The government appealed the decision, and the appeal went all the way to the Supreme Court. To avoid further expenses, the two sides agreed to an out-of-court settlement. In an interesting history-repeats-itself scenario, the caffeine content of a beverage is again attracting the spotlight. But this time it isn't Coke that is being targeted; the attack is against "energy drinks."

Energy drinks are very popular. There's a slew of them out there. They contain a variety of herbs, vitamins, usually loads of sugar, and always caffeine. A lot of caffeine—up to 500 milligrams per serving. Compare that with a maximum of 70 milligrams per serving for a cola-type beverage. The effect of this much caffeine on children and adolescents is unknown, but adverse effects on the developing heart and brain are a possibility. Many energy drinks also contain an extract of guarana, a plant that contains a significant amount of caffeine and adds to the total caffeine content. And then, of course, there is the sugar content. At least as much as in soft drinks. There's yet another effect: energy drinks will deplete your bank account. Why not just drink water?

In 1912, *The Strand Magazine* in Britain featured an ad for Sequarine, touted as the "Medicine of the Future." It was named after Charles-Édouard Brown-Séquard, the French physiologist who is considered one of the founders of modern endocriniology. What was the supposed "anti-aging" ingredient in Sequarine?

In 1889, at a meeting of the Société de Biologie in Paris, Brown-Séquard presented the results of a most unusual experiment. He described his feelings of rejuvenation after injecting himself with an aqueous extract made from the crushed testicles of young dogs and guinea pigs. Such extracts would eventually be marketed as

"Sequarine Serum," with claims of embodying "the very essence of animal energy." It is doubtful that Brown-Séquard had anything to do with this particular product, since he had passed away in 1894, long before Sequarine appeared on the market. However, it is known that, following publication of his work in *The Lancet* in 1889, Brown-Séquard had produced testicular extracts for use by physicians. He was not interested in profits, and he provided the extracts for free. All he asked was that doctors who used them report their experience to him.

Brown-Séquard believed that "infirmity in old age may in part be attributed to deteriorating functioning of the testicles." Apparently, he had been alerted to this possibility by his observation that eunuchs tended to be sickly and aged poorly. He attributed this to a reduction of some substance that was produced by the testes. It should be made clear that Brown-Séquard was no quack. He had forged a highly reputable career based on his studies of the nervous system. To this day, a certain type of damage to the spinal cord, resulting in paralysis and loss of sensation to touch on the same side as the injury and loss of sensation to pain and temperature on the other side, is known as Brown-Séquard syndrome. He had also postulated that certain substances could be released into the bloodstream that caused effects on distant organs. Indeed, he was on the track of hormones!

When he was seventy-two years old and feeling that he was growing increasingly tired and unable to spend the customary long hours in the laboratory, he decided to become his own guinea pig and test his ideas of rejuvenation. He would inject himself with testicular extracts. Since there was a dearth of human donors, Brown-Séquard decided on crushed animal testes. Dog testicles were first, but the supply ran out after a couple of days, and he switched to guinea pig organs. Before long, Brown-Séquard claimed that injecting a preparation of these animal testes led to renewed vigour. Once again he was able to dash up a flight of stairs and work late into the night. And his experience in the *pissoirs,* those famous French public urinals, was impressive. His urine stream, he claimed, reached some 25 per cent farther after his injections. Very interesting, especially considering that dogs have been known to project a pretty fair distance. Apparently, his chronic constipation also improved.

But what really caught the public's attention was Brown-Séquard's remark about other possible effects of testicular extract injections: "I might add that other powers, which admittedly hadn't deserted me entirely but were incontestably debilitated, have also improved markedly." The quacks were quick to capitalize on such claims and were soon promoting various elixirs and linking them to Brown-Séquard, much to his dismay. One can only guess at the number of men who were expecting a rise in their physiological fortunes but got a dose of blood poisoning instead.

Could it be that Brown-Séquard actually experienced a hormonal effect produced by testosterone? Not likely. It is known that the testosterone synthesized in the testes is quickly released into the bloodstream, and little remains behind. Indeed, this was confirmed in 2002 by a study published in *The Medical Journal of Australia.* Researchers duplicated Brown-Séquard's extraction and measured testosterone levels. The amount of testosterone was way below that expected to have any physiological effect. So it is a good bet that what the French physiologist experienced was the power of the placebo. He did note, interestingly, that the rejuvenation effect lasted only about four weeks. This did lead to others searching for longer-lasting effects, marking the beginning of research into hormonal therapy.

☆

What laboratory technique derives its name from the Greek for "coloured writing"?

Chromatography is one of the most common techniques used to separate compounds in the laboratory. Believe it or not, when you sniff the aroma of a cup of coffee, you are actually smelling a mixture of hundreds of compounds. How do we know that? Tip your hat to *chromatography*, a technique without which the practice of modern chemistry is almost inconceivable. Indeed, the separation of components of a mixture is one of the fundamental challenges a chemist faces. It is critical whether looking for bisphenol A in urine, a pesticide residue on an apple, or the active ingredient in a herbal remedy. Furthermore, a chemical synthesis almost always results in a mixture of compounds from which the desired ones have to be separated.

Chemists use a number of techniques to separate mixtures, including distillation, crystallization, filtration, centrifugation and solvent extraction. But chromatography heads the list. Russian botanist Mikhail Tsvet coined the term in 1901 from the Greek for "colour writing." Tsvet trickled an extract of green leaves through a glass tube filled with calcium carbonate. He was thrilled to see the originally-green solution separate into a series of coloured bands as each component pigment adhered to the calcium carbonate to a different extent. There was the orange band of carotene, the yellow of xanthophyll and a couple of green bands that Tsvet correctly concluded were due to chlorophylls that differed slightly in chemical structure. To the Russian botanist, it must have seemed as if coloured ink had been used to dye the white calcium carbonate, hence "coloured writing." Because of this classic experiment, Tsvet's name has been intimately associated with the invention of chromatography. But, as is almost always the case with inventions, the story is more complicated than that, and indeed, more colourful.

The *Annual Review of Progress in Chemistry*, published in German in 1862, some forty years before Tsvet's experiment, includes the following lines: "Friedrich Goppelsröder has shown that Schönbein's observation, whereby solutions of various substances are aspired with very different rates and intensities by filter paper, can be used to separate and distinguish different dyes contained in the solution." That is a clear description of what we now call *paper chromatography*, yet Goppelsröder and Schönbein hardly ever get a mention when it comes to the invention of the technique. Schönbein is certainly well recognized as the first chemist to make nitrocellulose, a discovery that led to the production of smokeless gunpowder. But his experiments with separating compounds using filter paper are largely forgotten.

In fact, Schönbein clearly showed that when filter paper was dipped into a water solution of acidified litmus, the water rose up the paper, eventually forming zones of different colour. But it was Goppelsröder who capitalized on Schönbein's remarkable separation: "I saw in these observations the key to a new analytical method and with M. Schönbein's permission I started to examine the behaviour of dyes from this aspect."

Goppelsröder went on to analyze various mixtures, including ones that were composed of colourless compounds. He developed methods to visualize these on the paper by spraying with various reagents that converted colourless compounds into coloured ones. This technique is still widely used today. Colourless amino acids, for example, develop a bluish tinge when reacted with ninhydrin.

Interestingly, in 1944, Archer Martin and Richard Synge received the Nobel Prize in chemistry for "the development of paper chromatography to characterize proteins by separation of their amino acid components." They are often referred to as the inventors of paper chromatography, which is clearly incorrect, seeing that Schönbein and Goppelsröder had used the technique some eighty years earlier. Indeed, Goppelsröder even separated biochemical substances such as those present in urine. He described analyzing 507 urine samples from 178 patients with 86 diseases in a valiant attempt to establish correlations.

But even Schönbein and Goppelsröder may not have been the true inventors of chromatography. There is evidence that, long before their pioneering work, dye makers checked the quality of their batches by dipping a white string into their vats. The different components of the dye mixture travelled up the string to different extents, leaving a range of coloured bands on the string.

Archer Martin, with colleague A.T. James, also developed *gas chromatography*, perhaps the most useful chromatographic technique. In 1944, Erika Cremer in Australia had shown that compounds that at room temperature were liquids could be separated from each other by converting them to gases, and then using an inert gas such as helium or nitrogen to push them through a column packed with some absorbent material. Based on this experiment, Martin and James developed the widely used technique of gas chromatography. Injecting a mixture into a gas chromatograph results in individual compounds emerging from the instrument at different times depending on how strongly the components bind to the packing material. The gas chromatograph is equipped with a detector that translates the emerging gases to a series of peaks on a moving chart paper. The number of peaks represents the number of components in the mixture. Coffee would yield hundreds.

The time it takes for a compound to travel through the column is characteristic of that compound. A gas chromatogram therefore can be used not only to determine the number of components in a mixture, but to identify specific compounds by comparing their "retention time" to that of known substances. Anytime you hear that chemists have found a certain phthalate in someone's blood, or a steroid in an athlete's urine, or some poison in a victim's food, it's a safe bet that gas chromatography has been at work.

In retrospect, the invention of chromatography was a team effort, although Goppelsröder should certainly get as much credit as Tsvet. Interestingly, Tsvet was of Russian origin but was educated in Switzerland, where Schönbein and Goppelsröder carried out their work decades before. Whether Tsvet was aware of their work is impossible to determine. As a final curiosity, Tsvet in Russian means "colour." Did he coin the term *chromatography* in a tongue-in-cheek fashion to commemorate his own name?

What gas was discovered because of a Scottish physician's interest in "magnesia alba"?

Carbon dioxide—or, as its discoverer, Joseph Black, called it, "fixed air." Black enrolled in medicine at Glasgow University in 1748, where he was smitten by the lectures of William Cullen. Although Cullen was professor of medicine, he had a year earlier instituted the first lectures in chemistry, a science he was convinced would prove to be of use for physicians. Black was intrigued by Cullen's chemical discussions and asked to be his laboratory assistant. It was here that Black began to work with "magnesia alba," a white mineral we know today as magnesium carbonate.

He noted that when the substance was heated it lost weight, and cleverly surmised that this was due to a gas that was given off, which he called "fixed air." That gas, of course, was carbon dioxide. The residue left behind after the release of the "fixed air" proved to be strongly alkaline when dissolved in water. This residue is magnesium oxide, which dissolves in water to form magnesium hydroxide.

Since, at the time, all medical students were required to write a thesis, Black had to look for some health connections. And he found them. Magnesium hydroxide was an effective laxative, as well as an effective antacid. It is used for these purposes to this day.

Black went on to become a professor of both medicine and chemistry at Glasgow and Edinburgh Universities, where the chemistry buildings were eventually named after him. His introduction of quantitative measurements came in handy, because his position as professor of chemistry at Edinburgh was unsalaried. As was the custom at the time, professors had to collect tuition directly from their students. This was a great driving force for producing popular lectures, which by all accounts Black managed to do. He was well known for his lecture demonstrations, many of which involved magnesia alba. To collect fees, Black would sit at the entrance to the auditorium, where students would hand him gold coins. But he soon discovered that some students were cheating by clipping gold off the coins. This is where the professor's previous experience with scales came in handy, and he began to weigh the coins he collected. History does not record what happened to the students who tried to scrimp on tuition.

Alongside his teaching, Joseph Black maintained an active medical practice, but he himself suffered from poor health, which some have interpreted as vitamin D deficiency, based on records that his health improved when he moved from the city to the countryside. Here, milk from cows fed on grass had more of the vitamin than milk from city cows raised in dark barns. Today, he would not have this problem, since milk is fortified with vitamin D.

☆

From 1804 to 1921, the DuPont Company hired hundreds of women to strip willow trees of their bark. Why?

DuPont was the prime producer of *black powder*, the gunpowder used for hunting, mining, construction and war. It was made by combining sulphur, saltpetre and charcoal. The purest charcoal was made from willow wood because it is very low in minerals. The demand was huge—DuPont needed some five million pounds of charcoal a year to make gunpowder and explosives.

Why did Spanish galleons transport large amounts of mercury to the New World after the sixteenth-century "Conquest"?

Silver was much prized in Europe, and the Spaniards were thrilled to have discovered that the New World offered rich sources of the metal. The Incas and Aztecs, like the ancients of the Old World, had developed methods of extracting silver from ores and had built up an abundant supply of the metal for ornamental use. The Spaniards robbed the natives of their precious metals and quickly discovered that there was more to be had where those came from. South America was rich in silver deposits.

Although the metal does occur to a small extent in its native state, most of it is found in ores in the form of silver sulphide, usually mixed with lead sulphide. The techniques to isolate and purify silver—basically smelting and cuppelation—had been developed independently in the Old World and in the Americas. In the Old World, smelting dates back some eight thousand years, while in the Americas there is evidence that pre-Inca civilizations had mastered the smelting of copper and silver at least six centuries before the first Europeans arrived in the sixteenth century.

The smelting of metals from their ores is one of the most important chemical processes ever developed. Metals can truly be thought of as the cornerstones of civilization. Most ores are compounds of the desired metal with oxygen (oxides), sulphur (sulphides) or carbon and oxygen (carbonates).

Smelting is basically a two-stage process in which the ground ores are first heated to a high temperature to drive off unwanted carbon or sulphur, converting metals to their oxides. These are then heated with a source of carbon, usually charcoal or coke. This combustion process produces carbon monoxide, which strips oxygen from the metal oxides to yield carbon dioxide and the free metal. In the case of silver, since its ore is usually found mixed with that of lead, the product of smelting is a mixture of the two metals. This is where the ancient technique of cuppelation comes in. It is based on the principle that precious metals do not react as readily with oxygen or other chemicals as do base metals such as lead. In silver extraction, careful heating converts the lead to lead oxide, which is denser than silver, and the precious metal is left floating on top.

This was essentially the method being used in South America when the conquistadores appeared. It was labour-intensive, but at first, that was not much of a problem for the Spaniards. They used Incan labour, and when that wasn't enough to meet the European demand for silver, they imported African slaves. As the high-grade ores began to be depleted, the Spaniards searched for improved extraction methods. The answer came from a successful merchant, Bartolomé de Medina, who had learned from a German craftsman recorded in history only as "Maestro Lorenzo" that silver could be extracted from ground ores using mercury and salt water.

This gave birth to what became known as the *patio process*, which involved spreading the crushed ore, mixed with salt and mercury, in an open enclosure known as a patio. After weeks of mixing with horse-drawn paddles and soaking in the sun, a complex reaction converted the silver compounds to metallic silver, which then dissolved in the mercury to form an *amalgam*. The mercury would then be driven off by heat, leaving silver behind. It also left numerous dead workers, poisoned by mercury, behind.

An expansion of mercury production was central to the expansion of silver production, which meant more poisonings. Mercury is mined in the form of mercury sulphide or cinnabar, and heating this ore to extract the mercury results in toxic mercury vapours. It is estimated that some eight million natives and slaves died as a result of working in South American silver mines to feed the Spanish monarch's hunger for silver.

There was hunger of another kind. Workers in the mines were fed almost exclusively on *chuno*, perhaps the world's first processed potato product. And where did this come from? The Spaniards encouraged the cultivation of the potato, a vegetable native to South America, and then forced the natives to pay taxes in the form of *chuno*. Indigenous South Americans had developed a method of storing potatoes for years as a safeguard against famine. In the Andes, temperatures commonly fall to below freezing at night. Potatoes were left out to freeze, and in the morning, when the sun warmed them up, their liquid content could be expressed by the simple process of trodding on them. This resulted in what could be called a "freeze-dried" potato that could then be stored in permanently frozen underground caves. Later, *chuno* would be ground into flour and baked into bread, or rehydrated to be mixed with vegetables to make a stew. It was valued so highly that it was even placed into graves to sustain the dead on their journey to the other world.

Without the potato feeding the workers, the Spaniards may not have been able to exploit the silver mines, and that could have changed history. It was the profits from silver imported from the Americas that funded the creation of the Spanish Armada that ruled the seas until defeated by the British in 1588. The Spaniards eventually suffered in other ways from their reliance on silver. Because of their large silver supplies, they could afford to buy goods from other countries, especially China, and so never developed a manufacturing industry of their own. When too much silver production elsewhere led to inflation, silver declined in value and the Spanish Empire could not sustain itself. One could say that the potato led to its decline.

In the first century AD, Roman emperor Nero issued a decree that gave the emperor the exclusive right to wear what?

According to Nero's edict, only the emperor of Rome was allowed to wear garments that were dyed with "royal purple." Conquering generals and senators were allowed purple stripes and trim, but purple togas were only for the emperor. By Nero's time, the use of royal purple, or "Tyrian purple," was well established, although it was almost always limited to dyeing fabrics worn by nobility or high priests. Even the Bible describes royal purple being used for the tabernacle curtains and high priests' vestments. The dye owed its importance both to its beautiful colour and the difficulty in producing it.

The source of the renowned colour was the secretion of certain marine molluscs, with about ten thousand of the creatures needed to produce a gram of dye. As early as the thirteenth century BC, the Phoenicians had established what amounts to a dye factory at

Sarepta, on the coast of Lebanon. It was likely man's first large-scale chemical industry, and undoubtedly a very smelly one. In all probability, the discovery of the dye was the result of these molluscs being eaten and their remnants discarded and left to decay in the sun. Their oozing juice slowly transformed to a spectacular purple colour. The secretions of the shellfish are actually colourless until disturbance of the tissues liberates enzymes that initiate a cavalcade of chemical reactions, eventually leading to colour formation. Under the influence of oxygen and light, the colourless compounds produce a mixture of chemicals that are responsible for the alluring purple hue, the main one being dibromoindigotin.

It wasn't until the early part of the twentieth century that the complex chemistry of royal purple was clarified. By that time, this natural dye had lost its importance, thanks to William Henry Perkin's landmark 1856 discovery. Engaged in a futile effort to produce quinine from coal tar, Perkin, barely eighteen at the time, accidentally synthesized the world's first synthetic dye. It was a brilliant purple colour that achieved fame as mauve.

Queen Victoria set off what came to be called the Mauve Decade when she wore a silk gown coloured with Perkin's novel dye to the Royal Exhibition of 1862. And then Napoleon III's empress, Eugénie, triggered true mauve madness by wearing purple gowns. The origin of the dye may have changed from molluscs to coal tar, but purple maintained its royal association.

TRICKS
OF THE TRADE

Shake a can of Guinness and you'll hear a widget rattling around inside. What is its purpose?

The widget helps produce a creamy head of the sort that consumers expect to see in a glass of Guinness draft. Aficionados will tell you that the frothy foam, or head, that floats on top of the liquid is a critical component of the beer-drinking experience. So, what makes for good head? The basic science is pretty straightforward.

Gases dissolved in the beer escape from solution as the external pressure is reduced, forming bubbles that rise to the surface, where they are trapped by the liquid. That's the foam! Its nature depends on the amount of gas dissolved and the specific chemical composition of the liquid. So-called "surface-active" compounds in the liquid determine how far a thin layer of the liquid can be stretched before bursting, a property critical to the extent of foam formation and its longevity.

In the case of beer, polypeptides from the malt are the most important surface-active compounds, and the dissolved gas is carbon dioxide formed by the action of brewer's yeast on carbohydrates in the malt. Sometimes the beer is fortified with extra carbon dioxide through the addition of pressurized carbon dioxide.

Guinness contains relatively little dissolved carbon dioxide, yet a glass of draft has a creamy, long-lasting head that, for many consumers, accounts for much of its appeal. There's some heady technology involved here, beginning with the addition of pressurized nitrogen gas as the beer is being dispensed from the keg. Nitrogen is soluble in water, but not to the same extent as carbon dioxide, meaning that when the pressure is reduced, it comes out of solution more readily. The emergence of nitrogen from solution is enhanced by increasing the surface area of the liquid in contact with the surrounding air, which is accomplished by passing the beer through a plate perforated with tiny holes. And presto, there's your thick, creamy head!

Much research has gone into mimicking this process in canned Guinness. And it has paid off. The secret to a great head in canned Guinness lies in a widget inserted into the can before it is sealed.

So, what is a widget? A hollow, polypropylene gadget, with no openings save for one tiny hole. Originally the widget was a sphere, but these days other shapes are used as well. The empty widget is placed in the can, and the can is then filled with beer, but just before the can is sealed, a shot of liquid nitrogen is added. As the liquid nitrogen evaporates, it forms a layer of pressurized nitrogen gas, some of which dissolves in the beer and some, along with beer, is forced into the widget. What we now have is a gizmo filled with beer under nitrogen pressure. When the can is opened, the contents of the widget blast out in response to the reduced external pressure. Dissolved nitrogen is liberated, microscopic bubbles materialize, and a head, much like that seen in Guinness on draft, appears.

The current state-of the-art widget is shaped like a little rocket, with the hole at its bottom. Why? The idea is that you can now drink from the can and get the draft-beer effect. Instead of the pressure being released from the widget all at once, a little is released each time the can is tilted for drinking, producing a head with each sip. Guinness reportedly spent thirteen million dollars to add that little bit of pleasure to our lives. Cheers!

A jet of flame erupts when a Gummi bear is dropped into a test tube in which a white, crystalline substance has been heated with a Bunsen burner. What is the substance in the test tube?

Potassium chlorate. This is a classic chemistry experiment that, when performed properly, can teach a great deal about the role of oxygen in combustion. Improper performance, however, can lead to quite a different educational effect.

When heated in the presence of a catalyst such as manganese dioxide, potassium chlorate decomposes to yield potassium chloride and oxygen. Combustion reactions require a fuel, a supply of oxygen and a source of heat to initiate the reaction. In this case, the fuel is the mix of sugar and gelatin in the Gummi bear, oxygen is provided by the decomposing potassium chlorate, and the required high temperature is achieved by heating with the burner. Once the combustion reaction starts, it becomes self-sustaining. The heat produced is sufficient to cause the continuous production of oxygen from decomposition of the potassium chlorate, which actually begins to melt. Flame and smoke spewing out of the test tube give the appearance of a rocket engine. Actually, the test tube really does become a scale model of a rocket engine.

All rocket engines are based on some sort of a combustion process using an internal supply of oxygen. And just as can happen with a real rocket, things can go askew with the scaled-down version. Oxygen trapped under the molten potassium chlorate can cause the test tube's contents to splatter violently. But that's not the worst of it. A test tube with too small an opening, using too much potassium chlorate—or using a fuel containing possible catalysts such as acids in sour candies—can result in an explosion. That's why this demonstration requires safety precautions and should never be performed without a protective shield. But this demonstration does have more than just a "gee-whiz" value.

Reactions that generate oxygen in this fashion are widely used in airliners, submarines and even the space station. In airplanes, canisters of sodium chlorate are stored either overhead or in the backs of seats, connected with tubing to oxygen masks. Depressurization of the cabin activates a striking pin that, by means of its mechanical action, triggers the explosion of a mix of lead styphnate and tetrazene in a percussion cap. The heat generated then initiates the decomposition of sodium chlorate to sodium chloride and oxygen.

Such chemical oxygen generators have to be surrounded by effective insulating materials because they can get hot enough to start a fire. Accidental activation of improperly shipped canisters is believed to have caused the fire that led to the crash of ValuJet Flight 592 in the Florida Everglades in 1996.

What chemical makes self-cleaning windows possible?

Titanium dioxide. Cleaning windows is not one of life's great pleasures. So how about a window that cleans itself? Sound like science fiction? Well, it isn't! Self-cleaning windows are here, thanks to recent developments in *photocatalysis* and *superhydrophilicity.* Tongue-twisting terms, but well worth exploring. Titanium dioxide is a naturally occurring pigment long used to impart a white colour to paints, inks, plastics, paper, ceramics, cosmetics and even food. When used as a pigment, titanium dioxide is inert, but when ground into extremely fine particles, it exhibits photocatalysis. This simply means that, when exposed to light—specifically, ultraviolet light in this case—it is capable of acting as a catalyst for various chemical reactions.

On exposure to ultraviolet light, titanium dioxide releases energetic electrons, which in turn react with oxygen and moisture in the surrounding air to form extremely reactive superoxide and hydroxyl free radicals. These immediately engage any organic compounds that happen to be around, breaking them down to carbon dioxide and water. Grime on windows is mostly a mix of organic compounds that can decompose in response to the photocatalytic activity of a thin layer of titanium dioxide applied to the glass.

But light does something else to titanium dioxide. It makes it *hydrophilic,* or "water loving." Normally, glass is *hydrophobic,* or "water hating," which is why water tends to form beads on its surface. When the beads drip, they leave telltale streaks. However, when titanium dioxide is applied to the surface, water, instead of beading, spreads out into an even, thin sheet. This means that when rain hits the window, it spreads perfectly across the surface and drains down evenly without leaving streaks. Presto! Any dirt that has resisted the photocatalytic effect of titanium dioxide is rinsed away.

It isn't only glass that can benefit from the free radical–generating capability of titanium dioxide. Cement can be kept from greying as well. Perhaps the best example of this technology is the Jubilee Church in Rome, completed in 2003. Made of self-cleaning concrete, it is designed to last for a thousand years, hopefully maintaining its white lustre. Airborne substances can be decomposed as well when they come in contact with a light-activated titanium dioxide surface. That's why titanium dioxide–coated tiles in toilets and operating rooms can destroy germs on contact and help fight the spread of disease. In Japan, coated pavement stones have been used to reduce the effects of air pollution. In one study, photocatalytic paving decomposed about 15 per cent of the nitrous oxide released by cars travelling on a roadway. Paints formulated with photocatalytic titanium dioxide can even battle indoor air pollutants such as smoke, formaldehyde and benzene.

Powdered titanium dioxide has been used to treat polluted water and to increase the shelf life of fruit by eliminating ethylene gas, which is responsible for fruit ripening. Those inventive Japanese have even come up with photocatalytic deodorant pantyhose! But the most significant application of this technology may turn out to be in cancer treatment. Injection of titanium dioxide into a tumour, followed by exposure to directed ultraviolet light, can generate free radicals capable of destroying the tumour without affecting surrounding tissue. It all sounds good, perhaps too good.

What happens when the titanium dioxide–treated products end up being discarded? Might the titanium dioxide find its way into the ecosystem and affect beneficial microbes? Nobody really knows. Indeed, we may see more clearly through self-cleaning windows, but the long-term consequences of the photocatalytic technology are harder to see. And one more point: only the outside of the self-cleaning windows is coated with the photocatalyst. So you will still need some vinegar and a dose of elbow grease for the inside.

What do iodine, silver nitrate, ninhydrin and cyanoacrylate have in common?

They are all used to detect latent fingerprints. Touch any surface, and unless you're wearing gloves you'll deposit a spectacular array of chemicals we call sweat. There are fats, amino acids, sugars, chlorides, sulphates and phosphates, along with urea, lactic acid, metal ions, creatinine, choline, ammonia and, of course, water. These chemicals collect in the ridges on our skin and, when deposited on a surface, leave behind a fingerprint. But the problem, as far as forensic scientists are concerned, is that the chemicals that make up sweat, and therefore fingerprints, are colourless.

A number of chemical visualization techniques that can bring out latent fingerprints exist, with iodine fuming being the oldest. Iodine is a brown solid at room temperature, but it readily sublimes when warmed. Sublimation is the direct transformation of a solid to a vapour without going through a liquid phase. The iodine vapours dissolve in the fatty components of the fingerprint, forming a brown image. The image will slowly disappear as the iodine evaporates, but can be fixed if sprayed with a dilute starch solution. Starch and iodine form a dark blue, stable complex.

Treatment with silver nitrate is another classic method. This time, it is the salt content of the fingerprint that makes the technique possible. Silver nitrate reacts with sodium chloride to form a deposit of silver chloride, which turns dark brown or black when exposed to ultraviolet light. This is the classic reaction that makes photography possible. Ninhydrin reacts specifically with amino acids, which are always present in sweat, to form a purple-blue product known as Ruhemann's purple. The suspect object is sprayed with a ninhydrin solution, and the latent print appears after several hours.

Cyanoacrylate visualization is the latest technique to reveal fingerprints. It uses, of all things, Krazy Glue. The amazing adhesive power of this glue is achieved by a polymerization process. A container of the glue houses small molecules known as cyanoacrylates that can link together to form a long molecular chain when exposed to moisture. This polymerization reaction takes place on the surface to which the glue is applied, triggered by atmospheric moisture. In the case of fingerprint visualization, the object in question is sealed in a container with a sample of the glue. The molecules of cyanoacrylate evaporate, fill the chamber, and then polymerize in the moist fingerprints, forming a white image.

Unfortunately, all these clever chemical techniques can be nullified by the wearing of gloves.

During the Second World War, German spy Oswald Job was caught with a hollow key containing lead nitrate. What was he using this chemical for?

To make a secret ink. A solution of lead nitrate is colourless and can be used as "invisible ink." When the paper is sprayed with a solution of sodium sulphide, the writing turns black due to the formation of insoluble lead sulphide. Apparently, Job, working as a German spy, had been making use of this chemistry to send secret messages. Born in England of German heritage, Job had moved to Paris as a young man. Like many others who possessed a British passport, he was arrested when the Germans entered Paris and was put in an internment camp. It was here that he became friendly with the German guards and was eventually recruited to spy for Germany.

Claiming he had escaped from a prison camp, Job showed up at the British embassy in Madrid and was repatriated to the U.K. Before long, British postal censors became aware of an unusual amount of mail from Job to other former inmates of the German detention camp, leading to his being questioned by the authorities. An officer noted that Job lived in a small room, but had an unusually large set of keys. It turned out some of these were hollow and had been designed to store chemicals. Job was accused of being a spy, but claimed that he had only pretended to collaborate with the Germans in order to escape from the prison camp. The judge didn't buy the story, and in 1944 Oswald Job was executed under the British Treachery Act, which had become law in 1940.

The lead nitrate–sodium sulphide reaction was by no means the only one used to transmit secret messages. Invisible lead nitrate markings could also be revealed by reaction with potassium iodide. Lead iodide, with a characteristic yellow colour, forms readily. Even simple household chemicals were sometimes used. Writing with an ammonia solution is colourless until it is exposed to red cabbage juice, which turns it blue. Then there is the old classic: lemon juice. Writing with it is invisible until the paper is exposed to heat, turning the writing brown. Apparently, this was a favourite during the French Revolution. When life serves you lemons, you can make lemonade—or, if needed, invisible ink.

In the "Satan's telegram" illusion, a magician is handed an envelope, opens it and begins to read the telegram inside. Then, suddenly, smoke comes from the telegram and it bursts into flames. What chemical was applied backstage to create this effect?

White phosphorus. Magical effects that rely on chemistry are rarely performed on stage because they require a degree of chemical expertise if they are to succeed and amaze. Most magicians, like most people, have a fear of "chemicals." But chemical magic can be truly spectacular when properly performed, and Satan's telegram is an excellent example. It requires the application of a few drops of a solution made by dissolving a piece of white phosphorus in carbon disulphide. When the solvent evaporates, the phosphorus is left on the paper to burst into flame as it reacts with the oxygen in the air.

Timing is critical to ensure that the paper catches fire at the right moment, coinciding with the magician's patter. Safety is an even more important consideration. The magician must handle the "telegram" very carefully, and great pains must be taken backstage to ensure the phosphorus solution doesn't spill.

The ease with which white phosphorus burns has made it useful for the production of tracer bullets and incendiary bombs. During World War II, thousands of phosphorus bombs were dropped on German cities such as Dresden and Hamburg, triggering devastating fires. Smoke grenades are also based on phosphorus, the white smoke being phosphorus pentoxide that forms as the element ignites.

Phosphorus was first isolated by the German alchemist Hennig Brand in 1669. Like many alchemists at the time, Brand was searching for the secret of life. He thought he might find it in urine. After all, it was gold coloured, and gold never tarnished, seemingly lasting forever. Perhaps, he reasoned, humans were born with a certain amount of life-sustaining gold in their system, which was slowly eliminated in urine.

One day, Brand decided to boil urine to drive off water and see what would be left behind. A residue remained on the side of Brand's flask that, upon heating, yielded a vapour that the alchemist condensed in water. When the water evaporated, Brand was stunned to find a deposit that glowed in the dark and then caught fire.

The alchemist likely thought he had discovered the secret to longevity, and maybe even the elusive "philosopher's stone," capable of turning base metals into gold. As it turned out, he hadn't discovered the secret of life, but he did make an important discovery. Brand was the first to isolate the element that was eventually named phosphorus from the Greek for "bearer of light." And it sure did bear light, as well as heat. Phosphorus matches, introduced in the early 1800s, easily beat flint stones for kindling a fire.

By this time, it was known that bones were made of calcium phosphate and that dissolving them in sulphuric acid would yield phosphoric acid. Heating this acid with charcoal yielded phosphorus. No longer was there any need to mess around with urine. Isolation of phosphorus became even easier in the early twentieth century, when large deposits of various phosphate minerals were discovered. Phosphorus was now readily isolated by heating the phosphate-bearing rocks with coke in an electric furnace. The phosphorus so formed has to be stored under water to prevent it from catching fire. Most phosphate rock, however, is not turned into phosphorus, but into phosphoric acid used for the production of phosphate fertilizer, without which the world cannot be fed.

Large amounts of phosphates are also used in the detergent industry to prevent calcium and magnesium, the "hardness" minerals in water, from interfering with the cleaning ability of the detergent. Unfortunately, excessive phosphate introduced into natural water systems can cause fertilization of plants and algae, which, on their death, use up the dissolved oxygen in water that is essential for the survival of fish. Hence the emergence of a number of "phosphate-free" cleaning products. Clearly, then, phosphorus has some properties that go well beyond producing magical effects on a magician's stage. When it comes to phosphorus, it seems all the world is a stage.

✬

It's made of cotton cloth coated on one side
with polyethylene and on the other with a rubber
adhesive. What is it?

Duct tape—the tape with a thousand and one uses. Some of them
decidedly unearthly. Remember "Houston, we have a problem"?
And how the *Apollo 13* astronauts were forced to use the lunar
lander as a lifeboat when an explosion ripped a hole in the side of
their command module on the way to the moon? The problem was
that the lunar lander was not equipped with a filtration system
capable of absorbing the carbon dioxide exhaled by three astronauts, forcing engineers in Houston to come up with a quick fix.
It seemed that the only chance was to fit the command module's
filters to the lunar lander's air system. Duct tape did the trick, and
saved the astronauts' lives. Later, on *Apollo 17*, duct tape was used
to mend the broken fender of the lunar rover, preventing the vehicle from being incapacitated by lunar dust.

On earth, duct tape has myriad uses, from holding broken tail
lights in place to securing wigs. A judge in Ohio ordered an
abusive defendant's mouth to be duct-taped shut during a
trial, and a much-publicized bit of research even suggested that
it was useful in the treatment of warts, although subsequent,
better-designed studies failed to corroborate this. Just about the
only application that duct tape is not great for is fixing ducts.
Researchers at the Lawrence Berkeley National Laboratory carried
out carefully controlled experiments demonstrating that duct
tape, defined as any fabric-based tape held in place by a rubber
adhesive, failed to seal ducts under realistic conditions. Indeed,
building codes generally call for the use of special metal tapes for
sealing ducts and prohibit the use of duct tape.

The tape was first made during the Second World War,
supposedly for making emergency repairs on the battlefield.
According to some accounts, because it was waterproof, it was

referred to as "duck" tape, since water rolled off it as it does from a duck's back. Another possibility is that the original fabric used in the tape was *duck cotton*, a type of canvas made by tightly weaving cotton fibres and deriving its name from the Dutch term *doek*, for linen canvas. In any case, the suggestion is that after the war, when the tape found widespread use in the reparation of air ducts, *duck* metamorphosed into *duct*. Some etymologists argue that there is no evidence that the word *duck* was ever used in connection with the tape, and that it derives its name from one of the earliest uses— which was, in fact, the sealing of ducts.

Today, there are many varieties of duct tape, but they are all made by coating cotton fabric on one side with polyethylene and on the other with some sort of rubber adhesive. A great deal of chemistry is involved in making the adhesive to ensure that it forms a strong bond but can still be removed when desired. Either natural or synthetic rubber is mixed with a petroleum-derived resin and some calcium carbonate to form the pressure-sensitive adhesive. MacGyver, the television hero who solved crime with brains, not brawn, was a fan of duct tape, never venturing out without some in his back pocket. But I suspect the robbery suspect who had his mouth taped shut in the courtroom was probably not appreciative of duct tape's fascinating history, nor of the sophisticated chemistry involved in its production.

What novelty item has ammonium sulphide as its active ingredient?

The stink bomb. Just crush the glass ampoule, and the dreadful aroma of hydrogen sulphide will send folks scurrying from the scene. That classic stench of rotten eggs is produced when the ammonium sulphide in the stink bomb breaks down to hydrogen sulphide and ammonia.

Commercially, this ammonium sulphide solution is prepared by the exact reversal of the "stink bomb" reaction. Hydrogen sulphide gas—either isolated from natural gas, where it is a common contaminant, or produced by reacting iron sulphide with hydrochloric acid—is bubbled into a solution of ammonia. The resulting ammonium sulphide can be stored in a suitably sealed container.

Ammonium sulphide is an unstable compound and decomposes into ammonia and hydrogen sulphide. An equilibrium is established between the hydrogen sulphide in the headspace of the container and that dissolved in the solution. As soon as the container is broken, the dissolved ammonium sulphide rapidly decomposes and liberates copious amounts of the smelly gas. Really, really stinky, but also dangerous. Hydrogen sulphide is one of the most toxic gases known, and can be lethal. Stink bombs, however, aren't likely to kill people, because not all that much gas is produced, and the gas quickly disperses. But attempts to make stink bombs at home can be potentially dangerous.

The classic method is to cut the heads of strike-anywhere matches and seal them in a bottle with a few millilitres of household ammonia. The match heads contain phosphorus sesquisulphide, a compound that decomposes in water to yield phosphoric acid and hydrogen sulphide. The hydrogen sulphide then reacts with ammonia, and after a couple of days, you have a potent solution of ammonium sulphide, ready to release hydrogen sulphide as soon as the bottle is opened. If this is done indoors under poor ventilation, enough gas may be released to cause toxicity. It's not a good idea to play around with hydrogen sulphide, especially because it quickly deadens the sense of smell, and its continuous presence may not be realized.

Another joke-shop item that releases hydrogen sulphide is the "fart bomb." This consists of a bag containing calcium polysulphide inside another bag filled with an acid solution. Squeezing pops the inner bag, causing the acid to react with the calcium polysulphide to generate hydrogen sulphide gas.

Stink bombs and fart bombs are the stuff of pranks, but stinky-gas production can have a serious side. Both the Israelis and the Americans have developed riot-control bombs that liberate a horrific smell, causing crowds to disperse. These have been termed "skunk bombs" because they are supposedly synthetic versions of the compounds released by skunks, which have been identified as (E)-2-butene-I-thiol and 3-methyl-I-butanethiol. The scent of those will stop people from rioting pretty quickly.

But the honour of being the stinkiest substance belongs to "U.S. Government Standard bathroom malodor." Why would someone want to create such a thing? It's used to test the efficacy of deodorizers and air fresheners. And what is in this horrific mix? Just dissolve some skatole—which is the delightful fragrance of feces—naphthalenethiol, thioglycolic acid, hexanoic acid, p-cresyl isovalerate and N-methylmorpholine in dipropylene glycol, and you have it.

Although it is pretty repugnant, the "bathroom malodor" may have been rivalled by a mixture developed during World War II by the U.S. military with the intriguing name of Who Me? It smelled like feces and was dispensed as a spray from an atomizer. It was to be used by French Resistance fighters who would sidle up to German officers and spray them in an attempt to embarrass them. It wasn't a great success, with the sprayer ending up as stinky as the sprayee.

CREATURES
GREAT
AND SMALL

Why are cockroaches crazy?

The word *cockroach* comes from the Spanish *cucaracha*—"crazy bug."
Why crazy? Because when they are surprised by a light being
turned on, they scamper away in a wild, crazy, zigzag pattern.

Cockroaches hate the light. In fact, they belong to the order of
insects called *Blattaria*, which means to "shun the light." There are
some four thousand different species of cockroach that have
roamed the world for 300 million years. The German cockroach,
Blattella germanica, is the number one pest in North America and the
one that people are most familiar with. It commonly infests homes,
squeezes through cracks and feeds on almost anything, from
human food to wood, from soap to glue.

Cockroaches live in places where food, water and shelter are
available. Hence, roaches are often found in the moist environments of kitchens and bathrooms. They can also swim and
can stay underwater for as long as ten minutes or rest in one
spot without moving for eighteen hours a day. They eat only at
night and are omnivorous, but can go a long time without food.
The more we learn about their diet, the more disgusting they

seem, since they really will eat anything, including animal feces.

Cockroaches often come in contact with drains and sewers, where they can pick up bacteria and viruses from human excrement that can survive in the cockroach's digestive tract for months. They can carry salmonella, staphylococcus and streptococcus bacteria as well as viruses such as polio—however, the extent to which they spread disease is controversial.

When it comes to asthma, though, they are a health hazard. Cockroach debris, such as shed cuticles, eggs, saliva and feces, contains antigens, which are proteins that trigger allergic reactions. As the antigen becomes airborne and is inhaled, the effects can include respiratory symptoms such as asthma. Asthma is on the increase globally, and recent studies show that inner-city minority kids are especially at risk. Socioeconomic factors such as access to education, health care, insurance and medication may play a role, but the prevalence of asthma in these populations is most likely a consequence of poor housing conditions. The problem may be further exacerbated as parents, threatened by the crime on the streets, keep their children indoors.

It is crucial, therefore, not only to call an exterminator, but also to remove the cockroach antigens from the dwelling. For this purpose, the Food and Drug Administration has developed a polyclonal antibody home-test kit that detects cockroach antigens, which can then be removed with common household cleaners. This is especially important since the absence of cockroaches in a dwelling does not guarantee the absence of cockroach antigens. In fact, the antigens may persist up to five years after the elimination of cockroaches.

✿

You can die just from picking this animal up.
What is it?

The South American "poison dart" frog. There are several species of such frogs that produce such powerful toxins that just brushing against their poisonous skins is enough to kill an adult human. One one-hundredth of a milligram of batrachotoxin—an incredibly small amount—is lethal.

The frogs are brilliantly coloured, a natural warning to predators. But it was undoubtedly their stunning appearance that first attracted South American natives who then learned through experience that handling the frogs was not a good idea. On the other hand, a touch of toxin on the tips of arrows or darts was enough to bring down large prey.

But how do you handle the frogs to extract the poison? Natives pick the creatures up with a leaf and then rub the tip of a dart in the poisonous mucus secreted by the frog to prepare it for the blowgun. Poison frogs are thought to obtain their toxicity from consuming poisonous insects. The frogs themselves have developed an immunity to the poison. Perhaps the most spectacular of the poison dart frogs is the stunningly blue *Dendrobates azureus*. But it is very much a case of "look, but don't touch."

Not many predators can withstand the poison of a poison arrow frog. *Leimadophis epinephelus* is one known snake that seems to be immune to most of this frog family's poison. Amazingly, there is a positive side to the powerful toxins that poison dart frogs excrete. *Epipedrobates tricolor* from Ecuador may eventually furnish us with a new painkiller that blocks pain more effectively than morphine. Epibatidine itself is too toxic to use, but chemists have determined its molecular structure and are synthesizing various derivatives that look to be hopeful. Morphine can suppress breathing and lead to constipation, but tests have shown that some synthetic derivatives of epibatidine may avoid these side effects. Obviously, toxins can be used to kill or to cure.

What does the African crested rat do that no other mammal is known to do?

Many animals have evolved various means of protecting themselves from predators. Porcupines deter attackers with their sharp quills; skunks spew their noxious spray. Monarch butterflies, "poison dart frogs" and the pitohui bird of New Guinea secrete potent toxins that teach a predator to never attempt to dine on such a creature again. The toxins in these instances are derived from the diet, likely from insects and plants that are consumed. The monarch butterfly, for example, produces compounds of the cardenolide family from milkweed, their prime food source. But the African crested rat is unique in that it is the only known mammal to actually apply a poison to its fur to protect itself from becoming a meal.

In this case, that poison is ouabain, a compound that is closely related to cardenolides. These are steroids found in a variety of plants, usually joined to sugar molecules to make up the *cardiac glycosides*, the most famous of which is digitalis. Ouabain is also such a cardiac glycoside. As drugs, cardiac glycosides can be used to improve heart function in congestive heart failure, but in the wrong dose they can also trigger a heart attack. That is just what can happen when a predator bites into an African crested rat.

This rodent can grow to be about twice the size of a sewer rat and is normally a pretty sluggish animal. But when threatened, it raises its back to part its mane, sort of like a porcupine, to reveal a series of long white and black hairs—which, unlike the porcupine's quills, are not sharp, but have been smeared with a poison. The rat's chemical weaponry can be traced to its habit of chewing on the bark of a tree known as *Acokanthera schimperi*, the very tree known for its use by African tribesmen to tip poison arrows. An arrow tipped with this poison can even bring down an elephant.

The rat mixes the chewed bark with its saliva and applies the mixture to the hidden short hairs on its back, ready to be exposed when a predator approaches. The poisoned hairs are different from other hairs in that they are more porous and have a special ability to pick up the applied saliva, with its content of cardiac glycosides. Videos of rats in captivity clearly show the process of chewing on a piece of bark and applying saliva to the shorter hairs revealed by a parted mane.

This elegant research was led by Jonathan Kingdon of Oxford University, who grew up in East Africa and was familiar with stories about dogs biting into crested rats and succumbing to their toxin. But if they happened to survive, the dogs learned a valuable lesson and never approached the furry creatures again.

An interesting question now arises about the rat's protective mechanism. Why is the rodent itself not poisoned by the tree bark's toxins? So far, nobody knows. But further research may reveal something about the mechanism of action of ouabain, a compound that is known to block the effect of a protein responsible for pumping sodium ions out of muscle cells. This can lead to the contraction of muscles with unusual force, which in the case of congestive heart disease can be beneficial, but in other instances can also trigger a fatal heart attack.

Conceivably, further investigation of the African crested rat's immunity to cardiac glycosides can cast new light on human heart function. In this case, we may want to give some further thought to the rat's ass, or at least to the fur that covers it.

Nobody has ever successfully swatted a baby housefly. Why not?

Because baby houseflies do not exist. A little backgrounder on the flies' reproductive cycle can clear up this seeming conundrum. Within a week of mating—a doggie-style activity lasting from a few seconds to several minutes (not unlike humans)—a female fly will lay an average of 120 eggs. Nice, warm manure is preferred. Then, in roughly twelve days, the eggs develop into larvae, then into maggots, and then into pupae from which flies emerge fully grown. The adult is then ready to flap its wings roughly three hundred times a second and begin its career of annoying humans in its continuous search for food, landing on our corn flakes, our steak and most disturbingly, on our skin. And the fly has some pretty disgusting dining habits. When it finds a source of nutrition, it regurgitates some previously digested food to help liquefy the fresh meal, which is then sucked up through its proboscis. But a little something is always left behind. And this can contain remnants of the places the fly previously visited, like a less than appetizing pile of horse dung.

Since a fly can harbour some 33 million bacteria internally, with another half-billion swarming all over its body, it isn't surprising that it can transmit diseases such as typhoid, cholera, dysentery, salmonella and even polio. Perhaps that's why President Obama took issue with one of these little disease-carriers alighting on his arm. His attack on the creature precipitated attacks from fly lovers. Nobody can claim that the President wouldn't hurt a fly.

Rae Thurston had much more patience with flies than Mr. Obama, but that arose out of necessity. Rae was married to Howard Thurston, one of America's greatest magicians, who during the early part of the twentieth century thrilled audiences with his "Wonder Show of the Universe." As a publicity stunt for a show in Florida, Thurston came up with an act that involved "hypnotizing" his wife and placing her in a sealed glass casket, which was lowered into a glass tank filled with water in the lobby of the theatre. She was to remain there for six hours as the public wandered by, amazed at her survival without a source of oxygen. Surely this was true magic!

Actually, the demonstration was entirely legitimate, with no trickery involved. Rae Thurston had learned to regulate her breathing carefully while remaining absolutely immobile. In this way, the air supply held out for hours. As a precaution, though, it was agreed that if something should go wrong, Mrs. Thurston was to wiggle her toe and some excuse would be found to lift her out without dispelling the illusion that she was hypnotized. And about two hours after the stunt had begun, an attendant noticed the toe signal and the casket was raised, with the public being told that a Hindu gentleman had protested that a Western person was using Yogic methods. A pretty lame excuse, but apparently the public bought it. When the glass casket was taken backstage, no problem could be noted. But when it was opened, a fly flew out!

The creature had been inadvertently locked in with Mrs. Thurston and had been tormenting the lady by walking all over her. She, not wanting to destroy the illusion, could not move to swat the intruder. The gallant lady lay there for over two hours, putting up with the torture to preserve her husband's reputation as a miracle worker. The organization known as PETA, People for the Ethical Treatment of Animals, would have been proud of her. After all, the fly escaped unharmed. They might have sent Mrs. Thurston a letter of commendation. President Obama, on the other hand, received a letter of reprimand for having dispensed with an innocent fly. I kid you not.

✿

Why would someone want to purchase bobcat urine?

Mousetraps are fine if you are dealing with a few rodents, but if there is a major infestation you need some heavy artillery. Bring on the bobcat pee!

Mice, rats, rabbits and moles are keen to avoid predators such as the bobcat. Scent plays a critical role. Predators usually mark their territory with urine, and the animals they prey upon learn to keep their distance from this fragrance. Most people who have a problem with mice don't have access to a bobcat that can prowl around and take care of pesky rodents, but anyone can purchase bobcat pee. Spraying this in appropriate places will send the rodents scurrying away.

So, what is it in bobcat pee that rodents can actually smell? Thanks to Harvard Medical School's Dr. Stephen Liberles, we now know. It's a compound called phenylethylamine.

Liberles found that a certain type of olfactory receptor in mice was strongly activated by bobcat urine. After separating the urine's components, he was able to trace the activity to one specific compound: phenylethylamine. An interesting question then arose: Was this compound particular to bobcats, or was it a universal predator product? There was only one way to find out—by collecting urine from as many different animals as possible and analyzing it for phenylethylamine content. So Liberles' students went off to zoos to collect output from leopards, jaguars and lions as well as from herbivores like cows, giraffes and zebras. Undoubtedly a challenging task, especially when it came to getting a giraffe to pee into a cup. Eventually, they managed to collect urine from thirty-eight species of predators and a variety of herbivores. All animals had some phenylethylamine in their urine, with carnivores having by far the most. Lions and tigers had three thousand times as much as the plant eaters.

To confirm that phenylethylamine was the active repellant, the researchers compared the effects of lion urine that contained the compound with urine that did not. This was quite easily done, since an enzyme known as monoamine oxidase readily destroys phenylethylamine and can be added to urine. Rats and mice were clearly deterred by regular lion urine, but not by the version that was devoid of phenylethylamine. There may be a whole new industry in the making here. Why bother buying real bobcat urine when a simple, readily made chemical can do the job? There's some experimentation needed to find out how well this works and what concentrations are needed, and whether it works for rodents other than mice. Maybe there's no need to buy wolf pee to deter coyotes and moose, fox pee against squirrels and Muscovy ducks, or coyote pee to fend off deer, raccoons and groundhogs. Maybe phenylethylamine will turn out to be one-stop shopping against all these varmints.

An interesting feature of phenylethylamine is that it is also found in humans. And we do not produce it as a warning to prey. It acts as a neurotransmitter, and there's some evidence that it plays a role in controlling mood. In fact, there is even a theory that it is the "chemical of love," because it is found in higher doses in people who are in the throes of this emotion.

Phenylethylamine is even found in chocolates, which has given rise to the often-told but incorrect story that chocolate can stimulate feelings of love. This is scientific folly, given that phenylethylamine is quickly metabolized by monoamine oxidase to phenylacetic acid. Phenylethylamine never makes it to the brain. In any case, the amount of this chemical in chocolate is trivial. It certainly doesn't keep mice away. I can attest to that. I've had mice in my office that have dined on chocolate bars.

✣

Viagra will supposedly save the rhinoceros from extinction. How?

The key word here is *supposedly.*

The little blue pill became available in 1998 and precipitated an onslaught of media accounts over the next few years. One that made the rounds was based on research published by Dr. Bill von Hippel, a psychologist from the University of New South Wales, and his brother, Dr. Frank von Hippel, a biologist from the University of Alaska. The rhinoceros, the stories implied, should be thankful for Viagra because Asian men were switching to the drug from rhinoceros horn, a traditional aphrodisiac for which rhinos were being hunted to extinction. But there were a couple of problems with the story.

Contrary to accounts that had circulated in the West, rhinoceros horn was never considered to be an aphrodisiac in traditional Chinese medicine, and the von Hippels had never suggested a connection to the rhinoceros. They had investigated the decline in sales of seal penises and reindeer antlers, both of which are traditional treatments for impotence. According to the von Hippels, sales plummeted after the introduction of Viagra, and surveys of Hong Kong apothecary shops confirmed that traditional Chinese medicines for impotence were taking a modest hit. That was interesting, because there had been no previous evidence of Chinese customers switching to Western medicines for other common ailments such as headaches or indigestion.

But the von Hippels suggested that failure to achieve an erection isn't comparable to a headache. Unlike with seal penises or deer antlers, the effects of Viagra are rapid and visible to the naked eye. The overall conclusion of the von Hippels' study was that Viagra might account for a reduced demand for several animal species that are overharvested for the purpose of treating impotence with traditional Chinese medicine. It was the press, not the scientists, who factored the rhinoceros into this equation.

There is an aspect of the rhinoceros connection that is correct. The animal is in danger of extinction, and there is a link to traditional Chinese medicine.

For over two thousand years, the horn of the rhinoceros has been used as a treatment for a host of ailments ranging from arthritis and fever to food poisoning and possession by the devil. Needless to say, long-term use does not prove efficacy. There is no evidence that rhinoceros horn is of any use to anyone except to its original owner. Drinking a brew made by adding powdered horn to water has never cured anyone of anything. That is not surprising given that the horn is made of keratin, the same protein found in animal hooves and our fingernails. Indeed, ingesting rhino horn has about the same therapeutic effect as chewing on fingernails. But it is considerably more expensive. A hundred grams of horn can sell on the black market for thousands of dollars. And prices have been increasing because treatment for cancer has been added to the rhino horn's miraculous effects after a story about a Vietnamese politician being cured had begun to circulate.

Almost all of the rhino horns being sold are the product of poaching in Africa. The South African government also issues a limited number of licences for "trophy hunting," but it appears that some of these licences have been issued to Vietnamese "hunters" who do not even know how to hold a rifle. The rhinos are actually being shot by their African guides, a practice that is expressly forbidden.

Instead of ending up on mantles, the horns are sold illegally on the street as medicine, for exorbitant amounts. The South African government tries to prevent this by putting a microchip into horns that have been obtained legally in order to trace them, but this is of questionable efficacy. Rangers also battle the poachers, with gun battles being a common occurrence. However, there are also reports of collaboration with the poachers in return for tidy payments.

So, no, Viagra is not going to save the rhinoceros. Not directly, anyway. But perhaps its efficacy compared with traditional Chinese medicines can be put to use. If the impotence of traditional medicines compared with scientifically researched drugs can be widely publicized, using Viagra as an example, perhaps a cloud can be cast over the use of rhino horn as a treatment in traditional medicine.

In 1981, the U.S. accused the Soviet Union of engaging in biological warfare in Laos, Cambodia and Afghanistan by dispensing what was termed "yellow rain." What did yellow rain eventually turn out to be?

Yellow rain was nothing other than the excrement of giant honeybees of the species *Apis dorsata.* The Cold War was still hot in the early 1980s, when Secretary of State Alexander Haig publicly accused the Soviet Union of supplying Vietnam and Laos with what he called "lethal chemical weapons." Haig claimed that low-flying aircraft had been seen spraying a yellow, oily liquid that caused seizures, blindness and internal bleeding among victims on the ground. The U.S., according to Haig, had physical evidence that a "yellow rain" composed of spores of a fungus known as fusarium had been dispensed over parts of Southeast Asia by the Soviets.

What was the physical evidence? Apparently, some yellow droplets had been sighted, and an analysis of one leaf and one twig had shown contamination with trace amounts of trichothecenes, fungal metabolites that supposedly did not occur naturally in that region of the world.

Trichothecenes are nasty compounds that interfere with protein synthesis in cells—about that, there is no doubt. They are produced by a number of moulds that can infect grains like wheat, oats or corn. Some of these, such as the trichothecenes produced by *Stachybotrys chartarum*, can become airborne and present a risk through inhalation. Stachybotrys can grow in damp indoor environments and has been known to cause health problems after flooding.

The U.S. alleged that trichothecenes do not occur in Southeast Asia in the amounts that had been detected, and therefore had to have been introduced by some outside agent. This claim was never substantiated by independent researchers engaged by the United Nations, who in fact pointed out that trichothecene-producing moulds had indeed been observed in Asia, so there was nothing remarkable about finding some of these chemicals. Supposed victims of chemical attack were also examined and found to actually be suffering from fungal skin infections. While trichothecenes were detected in the blood of some people, this was as likely in areas never exposed to yellow rain as elsewhere, and likely was the result of eating mycotoxin-contaminated food.

As far as the yellow rain went, well, there was nothing remarkable about that, either. It was the harmless fecal matter of honeybees. Harvard researcher Matthew Meselson collected "yellow rain drops" and showed that they consisted largely of pollen that was typical of the area. Furthermore, each drop collected from the same leaf contained a different mix of pollen grains, as would be expected if they came from different bees.

The U.S. could not contest this finding, but responded that the Soviets had added pollen to the chemical mix deliberately because pollen was easily inhaled and would ensure proper delivery of the toxins. Seems a little far-fetched, given that no trichothecenes were ever detected in the yellow drops. In any case, the current opinion is that the whole yellow-rain affair was propaganda unsubstantiated by facts. It seems to share some common features with the allegation that Iraq possessed weapons of mass destruction.

The U.S. has never retracted the accusation that the Soviets supplied their allies with trichothecene-based biological warfare agents. While there appears to be evidence that the Soviets carried out research on such toxins, as did the Americans, it is very unlikely that this ever amounted to anything practical. Trichothecenes can be absorbed through the skin, making them a possible weapon in theory, but these chemicals would be very difficult to produce in any quantity. And quantity would be needed. Roughly three tons would be needed per square mile to ensure an effective killing dose.

None of this is to suggest that such mycotoxins are not dangerous. Natural contamination of improperly stored food can have lethal effects. The Soviets only knew this too well. During the German onslaught in 1942 and 1943, grain remained in the fields over the winter and became contaminated with fusarium. It is estimated that as many as 100,000 people may have died from eating bread contaminated with the mould. Trichothecenes can induce alimentary toxic aleukia, a dramatic lowering of white blood cells that increases the susceptibility to infections. No chemical warfare or bee dung was involved.

Why did Chinese archaeologist Li Zheng-yu set fire to a heap of wolf dung in 2006?

He wanted to observe the smoke. Li was exploring the origin of an ancient Chinese expression describing the outbreak of war as "wolf smoke rising around us." That reference is to the ancient Chinese technique of lighting beacon fires that warned of the approach of an enemy. Traditionally, the fuel used was wolf dung.

The Great Wall of China, parts of which predate the Christian era, is lined with thousands of beacon towers. These served as a sort of early-warning system that relayed military intelligence via smoke signals during the day and torches at night. A message could be transmitted as far as a thousand kilometres in just a few hours. Not quite email speed, but still pretty impressive.

A description of burning wolf dung first appeared in Chinese literature in the ninth century, precipitating conjecture about the special properties of this excrement. By the beginnings of the Ming dynasty in the fourteenth century, the accepted explanation was that dried wolf dung was used because it produced smoke that rose vertically without being dispersed by the wind. Such smoke was ideal for military signalling because it was visible from a great distance. Or so the story went.

A rationale was needed for the supposed effect, and it was found in the intestines of the wolf. Unlike other animals, which had intestines characterized by numerous folds, those of the wolf were supposedly straight. By some bizarre logic, the excrement that exited the straight digestive tract was thought to produce smoke that went straight up. It is curious that Ming dynasty scientists, who were clever enough to figure out that the addition of saltpetre and sulphur to beacon fires produced smoke that was more visible, never carried out dissections to test the wolf intestine theory. As time passed, their undocumented assertions slowly transformed into "facts," and the notion that the ancient Chinese used wolf dung for sending smoke signals was parroted in many a "historical" account.

The notion about the geometry of wolf intestine was, of course, nonsense, but Li Zheng-yu wondered if there might be *something* noteworthy about the smoke produced by wolf dung. Wolves did roam the steppes of ancient China in large numbers, so their dung would have been available to serve as fuel.

Amazingly, remnants of ancient fuel piles near beacon towers have been preserved by the arid climate and have been examined by archaeologists. A wide variety of vegetable matter was found, with occasional bits of animal dung, but never that of the wolf. Of course, the possibility remained that wolf dung was used but left no remains. So Zheng-yu decided to put the matter to a proper scientific test. He burned wolf dung and compared the smoke to that produced by burning the poop of other animals. No difference!

We can admire the ingenuity of the ancient Chinese in setting up an elaborate network of smoke signals, but the notion that they used wolf dung specifically because of the special smoke it produces belongs in the annals of mythology, not history. But that certainly does not mean the effects of burning dung need no further study.

About three billion people around the world cook and heat their homes using open fires and inefficient stoves. They burn wood, coal or crop waste if these are available, but when not, they burn animal dung. In many developing countries, piles of drying dung cakes are a common sight. Unfortunately, when these or other biofuels are burned in open fires or primitive stoves not equipped with chimneys, they produce a variety of air pollutants. Estimates are that nearly two million people die prematurely every year from illnesses attributable to indoor pollution produced by burning solid fuels.

Such fires produce a great deal of carbon monoxide, a compound that reduces the oxygen-carrying capacity of the blood. The inefficient combustion also yields particulate matter that can penetrate deep into the lungs, impairing immune function. Ventilation is generally very poor, and in many dwellings levels of small-particulate matter due to indoor smoke can be a hundred times higher than what is considered acceptable. Exposure is particularly high among women, who do the cooking, and young children, who spend the most time indoors near the domestic hearth. Half of the deaths due to pneumonia in children under five in developing countries are attributed to indoor smoke.

What is the connection between an Olympic gold medal and an Italian frog?

Gold medals are actually made of silver electroplated with gold. The process involves the use of an electric current to deposit gold on a silver medal immersed in a solution of a soluble gold salt. The first such electroplating process used a battery invented by Alessandro Volta, who got the idea from Luigi Galvani's classic experiment with a frog. In 1792, Galvani was dissecting a frog on a brass table when the frog's leg began to quiver upon touching with an iron scalpel. Galvani concluded that there was electricity stored in the frog. Colleague Volta did not buy the theory, postulating instead that the electricity was in the metals and not in the frog, and proved his point by producing electricity using stacking discs of dissimilar metals separated by sheets soaked in brine. He had constructed the world's first battery.

Volta was professor of Natural Philosophy at the University of Pavia, where his friend and colleague Luigi Brugnatelli was chair of chemistry. Brugnatelli became interested in Volta's battery and began to wonder about the effects of passing a current through various chemical solutions. In 1803, he discovered that immersing a piece of silver attached to the negative pole of a voltaic pile in a solution prepared by dissolving gold in aqua regia (a mixture of nitric and hydrochloric acids) to which ammonia had been added resulted in a coating of gold forming on the silver. This was a fascinating observation, but not suitable for commercial processes.

During the next few decades, improved batteries were developed, and many gold solutions were tested in hopes of developing a commercially viable gold-plating process. The breakthrough finally came some thirty years after Brugnatelli's seminal experiment.

John Wright was a veterinary surgeon who enjoyed experimenting with electricity and happened to read a paper written in 1783 by Carl Wilhelm Scheele, the famous Swedish chemist. Scheele had described his research on a dye known as Prussian blue, discovering in the process that gold dissolved in a solution of potassium cyanide. He immediately tried electroplating metals using the gold cyanide solution and was rewarded with the deposition of a firmly adherent layer of gold. Not only was this the beginning of the gold-plating industry, it also introduced a method that would be extensively used for extracting gold from gold-bearing ores. A cyanide solution reacts with the ore to form soluble gold cyanide salts from which pure gold can be recovered by treatment with zinc.

Today's Olympic gold medals are made by electroplating silver with gold, using an updated version of Dr. Wright's method. The stipulation is that each medal must contain at least six grams of gold. And it all started with an Italian frog.

In 1973, Arabella and Anita blasted off from Cape Canaveral aboard a spacecraft destined to rendezvous with Skylab, the U.S. space station. What task were they to perform aboard Skylab?

Arabella and Anita were spiders. Scientists were interested in learning whether weightlessness would affect their ability to spin a web, and if so, how. At first, both spiders wove ragged webs in their specially constructed window frame, but within a couple of days they had become acclimatized and constructed normal-looking webs. The effectiveness of the webs could not be determined, because no flies had been taken along for the ride.

The results of the experiment were published in the *American Journal of Arachnology*. NASA now knew that spiders could adapt to space flight and spin proper webs. The utility of this information is questionable.

But spiders have provided some insight into the action of drugs. Back in 1948, Hans Peters, a German zoologist, became intrigued by spiderwebs and planned on filming the web-spinning process. The problem was that spiders normally spin in the middle of the night—not the best time for research activities. Peters had an idea: Why not treat spiders to some stimulants and prompt them to work during the day?

Since spiders apparently love sugar solution, research assistant Peter Witt was given the task of doping the solution with various drugs to encourage daytime activity. He tried morphine, strychnine and dextroamphetamine. Why morphine was included isn't clear since, if anything, it induces sleep. But strychnine and dextroamphetamine are certainly stimulants. But they didn't stimulate the spiders to abandon their nocturnal activity.

However, Witt did note that the webs spun under the influence of the drugs were extraordinary. Some were made extremely meticulously while others looked like the spiders were on, well, drugs. A question now sprang up: Could spiders be used to learn something about the effects of drugs?

Witt constructed a special windowpane to house his spiders, administered an array of drugs including LSD mescaline, caffeine, Valium and psilocybin, and photographed his drugged arachnids. The most beautiful webs were spun under the influence of cannabis, the most symmetric ones were produced by spiders on LSD, and the most irregular ones by spiders given caffeine. So, if you want your spiders to catch flies, keep them away from your coffee.

Witt's work triggered an idea in the minds of two Swiss psychiatrists who had been struggling to determine the causes of schizophrenia. They were aware of the fact that healthy people on drugs such as mescaline or LSD sometimes exhibited schizophrenia-like symptoms. Could it be, they wondered, that schizophrenics had an excess of some natural chemical in their body that caused the disease? If so, these compounds might be present in the urine. But how to detect them?

Peter Witt's work with spiders raised a possibility. Perhaps feeding concentrated urine from schizophrenics to spiders might induce them to spin webs that would somehow be meaningful.

Urine was collected from schizophrenic patients, while urine from the researchers, who presumably were not affected by any mental illness, served as a control. A problem quickly cropped up: while spiders readily slurped up sugar solutions doped with drugs, they did not care one bit for those flavoured with concentrated urine. After one taste, they had enough and attempted to rub all vestiges of the solution off their mouths. They were traumatized enough by the experience to spin webs that differed from normal ones, but alas, there was no difference in the webs spun by the spiders given schizophrenic urine and those treated to the researchers' output. Spiders apparently cannot diagnose schizophrenia.

In case anyone is interested, spiders can spin webs even if two of their legs are missing. Groundbreaking work by a German biologist, Margrit Jacobi-Kleemann, in 1953 revealed that spiders with a leg removed from each side still spun normal webs. There is no need to conduct further experiments along these lines.

 CLAPTRAP

What is the origin of the word *claptrap*?

Back in the eighteenth century, a theatrical line that was delivered to shamelessly elicit, or "trap," applause from the audience was referred to as claptrap. It usually didn't have much meaning and became synonymous with nonsense. It is best to illustrate with an example. Here we go.

Thanks to chemical ingenuity, we lead a colourful life. Synthetic dyes have served up a feast for the eyes, but they may leave us starving for good health. Our reliance on these chemicals exposes us to a host of unnatural wavelengths that can affect our body chemistry. Until about 150 years ago, we had no choice but to rely on natural dyes. If you wanted red, you had better know where to find a madder plant. Blue came from indigo, purple from a type of mollusc. Why the different colours? Let's just cast a little light on the nature of light.

Light from the sun is composed of all the colours of the rainbow, as is readily demonstrated by passing it through a prism. Each colour represents a different wavelength, and since the different wavelengths are bent to a different extent as they pass through the prism, we see a rainbow of colours. The colour of any substance is determined by which of these wavelengths are absorbed and which are reflected. Madder root extract is red because the alizarin it contains reflects wavelengths that correspond to red and absorbs all other wavelengths.

Throughout most of our history, we were exposed to only natural colours—in other words, a mix of natural wavelengths. All of that changed in 1856, with William Henry Perkin's breakthrough discovery of synthetic dyes. Engaged in a futile attempt to make quinine, the young fledgling chemist accidentally discovered that chemicals in coal tar could be converted to dyes that had never been seen before. The mauve he produced took the world by storm, especially when it gained favour with Queen Victoria.

Soon, a variety of other dyes were produced from coal tar, and human eyes had to confront wavelengths for which evolution had not prepared them. Why should that matter? Because the eyes are windows to the brain, and the brain controls our biology. Depression, for example, can be triggered by reduced exposure to daylight, and flashing lights are known to cause seizures. Remember that we actually see with our brain; the eyes are only messengers. A stroke can completely destroy vision, even though the eyes still function perfectly well.

Our brain also controls the production of certain hormones and neurotransmitters such as melatonin, serotonin and dopamine. Simplistically, one can say that "rest and repair" hormones are activated by a lack of light, while "coping with stress" hormones are produced when light is abundant. But the situation is actually more complicated. Hormonal activity is governed by messages sent from the eye in response to specific wavelengths, and problems arise if there is exposure to an unnatural mix of wavelengths, termed the *unnatural wavelength effect* (UWE). This is characterized by an imbalance between "day hormones," like cortisol and insulin, and the "night hormones," such as the antioxidant melatonin and the immune system—enhancing prolactin.

The major problem, however, is that the unnatural wavelengths trigger the production of the recently identified compound esnesnon, a chemical that in the laboratory has been shown to have a yen for certain receptors on the surface of brain cells that have accordingly been christened yenolab receptors. Stimulation of these receptors has been linked to impaired central nervous system activity, which in turn affects all body functions. Because virtually all the dyes used today are synthetic, we are at increased risk for UWE. Whether we are gazing at clothes, furniture, our lacquered fingernails or our bedsheets, we are exposed to a range of wavelengths with which our ancestors never had to contend.

Luckily, there is a way to overcome the UWE. Exposing the eyes to an abundance of natural wavelengths can curb the production of esnesnon and reduce the chance of illness. Fluitex (flower and fruit exposure) therapy involves placing a variety of coloured flowers and fruits around the house to ensure our eyes are constantly exposed to all the natural wavelengths in the 600–700 nanometre range, effectively blocking the light waves reflected from synthetic dyes.

Subjects undergoing fluitex therapy report an enhanced feeling of well-being, a resolution of aches and pains and a brighter outlook on life. The effects appear to be directly proportional to the variety of colours displayed, with a spectrum provided by red roses, yellow daisies, blueberries, green apples, purple plums and oranges being especially therapeutic.

Fluitex therapy is devoid of side effects. It is also devoid of any sense. As is esnesnon. Just spell it backwards. Same with yenolab receptors. I made it all up. Did I snare anyone? Probably.

So, what's the point? That total nonsense can be made to sound totally believable. I know, because I get questions all the time asking if I think this or that bit of poppycock is true. For example, recently I was forwarded an email that's going around claiming that Wi-Fi is dangerous because "manmade wireless signals in the microwave range are unnatural and therefore incompatible with life." Just as much nonsense as my esnesnon. How about golf gloves equipped with a wristband to "produce negative ions to help your body's performance and recovery"? Bunk! Ditto for those "energy bracelets" that contain a hologram "embedded with frequencies that react positively with your body's energy field to improve your balance, strength and flexibility." You need a flexible mind to swallow that claptrap.

Energy bracelets are sort of benign nonsense. But matters become more serious when claims are made about some liquid preparation that purports to cure cancer by "lowering the voltage of the cell structure by about 20 per cent," causing cancer cells to "digest" and be replaced with normal cells. Might such poppycock not distract some desperate patients from proper treatment? How about the Quantum Xrroid Consciousness Interface Machine, "the most advanced medical assessment and therapy device in the world today"? What does it do? Well, it "loops all 200 trillion human cells within a 55-channel biofeedback system to gather bioenergetic data at nanosecond speeds, creating optical wellness." Are not some people bamboozled by such hogwash? Judging by the number of QXCI machines sold, they sure are.

And should you want to be in tune with even more nonsense, you can tap a "quartz crystal singing bowl" with a mallet, which will then "transmit energy into the atmosphere, filling your aura with vibrational radiance, which translates into seven main colours of the rainbow." You should also know that "through pure tone, one can repattern the energy field organization that ultimately affects the cellular expression of disease and wellness."

If that doesn't work, there are always "garments powered with infrared technology to enhance and improve athletic performance, recovery and general well-being." If you want infrared technology, just wear clothes. Any clothes. Infrared rays are nothing more than radiated heat. But maybe you should make sure these infrared garments, some of which also emit health-enhancing "minus ions," are not coloured with synthetic dyes.

What former actress is known for her support of "bioidentical hormones"?

Suzanne Somers is a former actress with a pretty face and probably firm thighs. After all, she did advertise the ThighMaster for years. As far as her smooth skin goes, she offers an electrifying explanation: she uses a gizmo that sends tiny jolts of electricity to give her facial muscles a "workout." Right.

Unfortunately, Somers has had to deal with problems that go beyond her thighs and wrinkles. She has had a bout with breast cancer. But neither her good looks nor her struggles with illness qualify her for donning the mantle of a scientific guru. Yet, that is just what she has become. And guruhood means that Somers influences many lives as she sounds off on diets, "bio-identical hormones," nutritional supplements and "alternative" cancer treatments. Her success in beating breast cancer, she claims, is linked to injections of mistletoe extract. Never mind that she had a lumpectomy and radiation treatment.

In Somers's book *Knockout*, some physicians—Dr. Nicholas Gonzalez, for example—are placed on a pedestal because they are "curing cancer" through unconventional means. Except that the "cures" are not supported by facts. Gonzalez's regimen of numerous dietary supplements and coffee enemas has actually been tested by the National Center for Alternative and Complementary Medicine, an organization not exactly adverse to alternative therapies. Patients fared more poorly with the Gonzalez treatment than with conventional chemotherapy. But for Somers, this doctor, who has been reprimanded by the New York State medical board for "departing from accepted practice," who was forced to submit to psychological examinations and undergo retraining, and who has lost malpractice suits in which he was accused of negligence in cancer treatment, is a hero to whom we should listen.

In Somers's follow-up *Sexy Forever*, she expands her scientific expertise to gene expression. In an interview publicizing the book, she says: "We have five cancer protective genetic switches in our bodies that get turned off by diet and lifestyle. One is turned off by toxins and chemicals; one by poor quality food, i.e., non-organic, pesticide-laden food; one by lack of sleep; one by stress; and one by imbalanced hormones. Now, what women really need to understand is that first and foremost, to turn back on your protective genetic switches you've got to get the hormone switch turned back on. Imbalanced hormones are a big factor in why women are fat, and when women get fat they get very unhappy." Why is such pseudoscientific blather given talk-show airtime?

Somers's scientific expertise is demonstrated by another example. When her oncologist asked why she was taking steroids, her answer was: "Who, me? Never!" It seems she had no idea that the bioidentical hormones she was taking were steroids. And let's not even mention the folly of self-treatment of an estrogen receptor—positive form of breast cancer with estrogenic hormones, which is exactly what Somers's bioidentical hormones are.

Why, then, do people regard her as an authority on beauty, weight loss and health? Because she is a celebrity. Talk-show sets have become lecture rooms, and books written by celebrities are the new texts.

You've probably not heard of Alex Reid. This martial arts fighter and actor is a big name in Britain. He makes the rounds of the talk shows, dates glamorous models, has been in movies and soap operas, and has even had his own TV show. How does he prepare for fights? Let's listen to his philosophy: "It's actually very good for a man to have unprotected sex as long as he doesn't ejaculate because I believe that all the semen has a lot of nutrition. A tablespoon of semen has your equivalent of steak, eggs, lemons and oranges. I am reabsorbing it into my body and it makes me go *raaaachh!*" Makes *me* go nuts.

Let's not even address the physiological and nutritional nonsense, but the bit about unprotected sex is dangerous claptrap. Speaking of which, here's British actress Julia Sawalha on travelling to places where malaria is endemic: "I don't get inoculations or take anti-malaria tablets, I take the homeopathic alternative called 'nosodes' and I'm the only one who never comes down with anything." Lucky Julia. Perhaps she can point us towards some trials that demonstrate how preparations with no measurable active ingredients—which is the case for homeopathic "drugs"—can prevent malaria.

Demi Moore also has views on health. She thinks it can be optimized by treatment with leeches. "They have a little enzyme and when they are biting down on you, it gets released in your blood and generally you bleed for quite a bit and then your health is optimized—it detoxifies your blood." Ridiculous. So is Tom Cruise's declaration that "psychiatry doesn't work; when you study the effects, it's a crime against humanity." According to Scientology, which Cruise espouses, our problems are caused by mental implants we received from space aliens and should be treated by Scientology's mysterious "auditing" methods. So, who is committing the crime against humanity?

Professor Sarah Palin lectures on research funding. "Sometimes research dollars go to projects that have little or nothing to do with the public good. Things like fruit fly research. I kid you not." Not for the first time, she here shows breathtaking ignorance. Fruit flies are excellent models for the links between genes and disease and have provided important clues for exploring numerous conditions, ranging from birth defects to autism.

Let's conclude with model Heather Mills, who gives us this insight into the non-vegetarian's fate: "Meat sits in your colon for forty years and putrefies and eventually gives you the illness you die of. And that is a fact." No, Heather, it isn't.

What gemstone derives its name from the Greek for "wine" because of a belief in its ability to ward off drunkenness?

Amethyst is a beautiful gem that can exhibit a range of colours, with the most common one being a purplish hue that resembles the colour of wine. The notion that amethyst is an antidote for intoxication is, of course, a myth, but the ancient Greeks were deeply into mythology. Life was thought to be controlled by their many gods, each of whom had specialties. Dionysus was the god of wine and intoxication. And what did he have to do with amethyst? Here the story becomes complicated.

It seems that Dionysus's affections were not limited to wine; he had a taste for mortal ladies as well. A particular subject of his affections was a maiden who happened to be named Amethystos. Since the ancient Greek word for drunkenness is *methyl,* and the prefix *a-* can be translated as "not," one might surmise that Amethystos was not into the partaking of wine. And, it seems, neither was she into trysts with wine-loving deities. The girl presented a challenge, which is what might have excited Dionysus. She, however, was not interested in his advances and prayed to the goddess Artemis to remain chaste.

One of Artemis's specialties was virginity, but she also seems to have been blessed with somewhat of a whimsical nature. Artemis answered Amethystos's prayer by turning her into a white stone. That put a quick end to Dionysus's chase, at which point the irritated god poured wine on the white stone, dyeing the crystals purple. The purple stone became known as *amethyst* after the name of the maiden who had successfully resisted Dionysus's amorous advances by turning into stone. One wonders if she would have reconsidered her prayers had she known the outcome. Perhaps a little adventure with the god of wine might have been preferable to being turned into stone.

In any case, the myth of Amethystos was transformed into the belief that amethyst had the ability to counter the effects of alcohol intoxication, and the Greeks, as well as, later, the Romans, made cups out of the stone. The belief was that wine consumed from these cups would not lead to intoxication. Just to be on the safe side, the alcohol-neutralizing power of the cups was boosted by wearing amulets made of amethyst. It seems, though, that the gemstone didn't have much of an impact on the legendary drunken feasts of the ancient Greeks and Romans.

Chemically speaking, this is not surprising. Amethyst is just a form of quartz, which is basically silicon dioxide. It owes its colour to impurities, mostly small amounts of manganese or iron compounds. Amethyst is an inert substance that does not react with alcohol in any fashion. In spite of this, it is considered to be the stone of abstinence and is often present in the rings worn by bishops and cardinals.

In recent years, the nonsense about amethyst offering protection against drunkenness has been replaced by some more modern claptrap. Some inventive marketers now sell "metaphysical jewellery." One beguiling item is made from a mineral that features whitish veins, supposedly the result of having captured lightning. We're told that it can direct energy towards the fortification of one's weaknesses and create instantly noticeable healing. You can also purchase a version of amethyst called Merlinite, a name that alludes to its claimed magical properties. It is a remarkable gem indeed. Meditating with Merlinite can assist one in contacting souls of the deceased who wish to give a message to the living. It is also said to open psychic channels for intuitive understanding and to attract teachers from other planes to assist in one's studies during the dream state and in meditation.

And leave it to the quacks to keep up with modern technology. Instead of hawking amethyst as an antidote to alcohol, they now promote it as a substance that can protect against electromagnetic radiation from devices such as cell phones. The Bioelectric Shield, a pendant containing amethyst along with other minerals in a specific pattern, will shield the wearer from harmful radio frequencies, such as those generated by cell phones. As an added feature, it will also enhance personal energy. It does come with a reasonable disclaimer. Some people, the ad cautions, feel a difference, others feel nothing. No doubt.

What is the importance of bio-based succinic acid?

Succinic acid is used in the production of a wide variety of consumer items. That doesn't sound like claptrap, and it isn't. But we will get around to the claptrap part. First, though, a little "green" chemistry. In 2011, a U.S. Presidential Green Chemistry Challenge Award went to BioAmber Inc. for its production of biobased succinic acid! Admittedly this award doesn't quite have the aura of an Oscar, but for those in the know, it's a biggie. Why? Because it is given for an innovation that is likely to make the world a better place—a greener place.

"Green chemistry" is all about using renewable resources and designing products and processes that minimize the use of and generation of hazardous substances. So why is a novel process to produce succinic acid deserving of a "green" award? Let's see.

Chances are you have never heard of succinic acid, but in the chemical industry it is a prized commodity. Plastics, synthetic fibres, detergents deicing agents and plasticizers can all be made from succinic acid. It is also an approved flavouring agent for food and finds numerous applications in the pharmaceutical industry. Today, most succinic acid is made by reacting maleic anhydride with hydrogen. But therein lies a problem: the usual starting materials for the synthesis of maleic anhydride are either benzene or butane, both of which come from petroleum. That's why any process that can produce succinic acid from a renewable resource qualifies as a move towards greener chemistry.

Microorganisms such as bacteria or yeasts are like little chemical factories. Lactobacillus bacteria can make yogurt for us, and enzymes in yeast convert sugars into alcohol. Evidently, there are also microbes that can convert plant materials into succinic acid. As one might expect, the exact nature of these microbes is a trade secret, because biobased succinic acid is a potential gold mine. Just to give you an idea the prime target for replacement with biobased succinic acid would be petroleum-derived maleic anhydride, which now has a global market of some 1.65 million tons a year. Estimates are that succinic acid made from a renewable resource could bring in about $1.3 billion a year. BioAmber is not there yet. But it is running the world's first biobased succinic acid plant in France and is building a larger-scale plant in the U.S. Right now, the feedstock is glucose derived from wheat, but the goal for the future is to use agricultural and forestry waste.

Now for an intriguing question: Why does a green company have *amber* in its name? Amber may be yellow, orange or reddish, but green? The name, though, is an appropriate one. Amber was the original source of succinic acid! Indeed, the name *succinic acid* derives from the Latin term *succinum*, for amber, the hardened form of the liquidy resin that certain trees produce to ward off insects and to seal wounds. But should you see some resin oozing from a tree, don't bother waiting around for it to form amber unless you have a few million years to spare. That's how long it takes for the complex mixtures of chemicals in the resin to harden into amber.

The major components of the resin are compounds called terpenes. Over many, many years, they engage in a reaction to produce long molecules that eventually cross-link and form a tough three-dimensional network. Amber basically consists of these *polymerized* terpenes, with the colour depending on the extent of polymerization, as well as on air bubbles and non-volatile compounds that get embedded. Succinic acid is one of these non-volatile components, making up between 3 and 8 per cent of a sample of amber by weight.

Woe to any insect with the misfortune of getting in the way of the oozing resin. Its destiny was to become a mummified "inclusion" in the amber. Indeed, the most valuable amber pieces are the ones with spectacularly preserved insect inclusions. I have one of those. At least, I think I have one. I bought it in the Dominican Republic, which along with the Baltic countries, is a hotbed for amber. The reason I say I *think* I have one is that, thanks to the wonders of chemistry, fakes have widely invaded the world of amber. Celluloid, phenol-formaldehyde resins (Bakelite), polymethylmethacrylate (Plexiglas) and polystyrene have all been used to produce forgeries.

Supposedly, one way to tell the difference is by density. Amber will float in salt water, while the plastics will usually sink. Not a totally trustworthy method, because the inclusion of some air bubbles can allow plastics to float. A better test is to apply the point of a very hot needle to the sample. True amber will release a piney smell, something like the odour of turpentine. That comes as no surprise, since turpentine is produced from pine wood, and, like amber, is formed from terpenes.

Now for the claptrap. One day on my radio show, I mentioned that I was reluctant to risk damaging my amber piece by applying the hot needle treatment. I got a quick call from a listener advising me that there was a simple non-destructive way to tell if I had authentic amber. I just needed to get my hands on a teething baby. Before specifics, I was treated to a preamble.

Hippocrates used crushed amber to treat "dimming eyesight," I was told. I also learned that, during the Middle Ages, fumes from burning amber were used to ward off the plague, and that Martin Luther carried amber to protect him from kidney stones. And I was also reminded of succinic acid's role in the Krebs cycle, the set of biochemical reactions cells use to produce energy. All of this was supposed to convince me of the wondrous healing properties of amber. But for real proof, I needed that teething baby.

An amber necklace, around the baby's neck, my correspondent informed me, would get rid of the pain! No more screaming infant! (How I was to make a necklace without harming my amber was not clear.) The magical effect supposedly was due to the release of succinic acid, which then would be absorbed through the baby's skin. Actually, what we have here is magical thinking. Amber does not release any appreciable succinic acid by dermal contact, and there is absolutely no evidence that succinic acid has any painkilling effect. Of course, not having ready access to teething babies, I could not put the claim to a test. But why the amber? Why not just apply succinic acid to the skin? I'm sure that BioAmber would gladly furnish some for research purposes. After all, happy babies would certainly make the world a better place.

When I expressed my skepticism about the healing properties of amber necklaces on the air, I got another edifying call. This time, I was told that my skepticism was well founded. Anyone who bought into the amber necklace idea was being fooled. Only necklaces made of authentic Quebec hazelwood were effective for teething pain. *Ay yai yai.*

What is Pimat?

Chances are you have never heard of this special piece of fabric. But it does give me a chance to spin a good yarn. Over the years, a variety of exuberant and hopeful marketers have trooped into either my office or living room with all sorts of newfangled cleaning products, dietary supplements, water filters, air purifiers and cure-alls. I've listened to countless sales pitches for smell-absorbing seat cushions, gopher deterrents, healing pendulums, negative-ion generators, energizing bracelets, radiation neutraliz-ing pendants, magnetic shoe inserts and even crocodile traps. But perhaps the most memorable health aid I've ever been asked to examine was Pimat. What, you are asking, is Pimat? Hang on, it's going to be a bumpy ride.

My story begins with a call from two young women who asked if I would be interested in taking a look at a "revolutionary, cutting-edge, breakthrough product" they had come across on a recent trip to Poland. It was the answer to sleep disorders, headaches, depression, back pain, asthma, arthritis, impotence and infertility. I mentioned that I wasn't in the market for the latter two, but that I did sometimes have insomnia thinking about the numerous scam products promoted by con artists. They agreed that there were a lot of snake oil salesmen out there, but they had found something that really works. They knew this because they had tried it themselves and felt energized and mentally clarified! Sounds great, I said. More energy and mental clarity are always welcome. Bring on the Pimat! And they did.

The wondrous discovery turned out to be a piece of cloth, about 18 centimetres square, adorned with an asymmetrical pattern of ten red dots. I stared in amazement at the fabric that was to be the answer to so many of our health problems.

Pimat, I learned, stands for "pyramid mat," and "reproduces, in two-dimensional form, the healing power of the pyramids." The pattern of red dots "rebalances and restores our energy fields, our aura, because it generates an entire spectrum of radiesthetic colours essential for the body to maintain a healthy condition." *Ohhh...kayyy.* And how do you avail yourself of the powers of Pimat? Simple: you put it under your bedsheet and you sleep on it.

What scientific evidence is there, I sheepishly asked, that the health of our aura is being restored by this wondrous piece of spotted fabric? A hand quickly went into the bag and emerged with a 184-page treatise on this magnificent "regenerator of stamina." On the first page, I was introduced to the inventor, Ryszard Olszak, from whom I learned that the action of this product is based on the energetic effect of its geometrical configuration, or "neoenergy."

He does admit that "in spite of much research, the nature of this phenomenon could not be explained in terms of physics, and the energy emitted has not yet been measured by any conventional methods." Yet, he knows it exists. In order to characterize neoenergy, I was informed, "radiesthetic colours are used." What are these? Let's go to the horse's (or perhaps in this case, ass's) mouth.

"The aura is composed of all the colours of the spectrum, characterized by different wavelengths corresponding to colours, known as 'radiesthetic colours.' If one is ill or tired [off-colour!], there will be colours missing or faded or holes in the aura." It seems that in 1991, Ryszard Olszak discovered, through radionics, a pattern that generates an energy called neoenergy, or the energy of shape.

This produces all the colours of the spectrum and restores any that are missing." There you have it. Convincing rationale as to why that Pimat works.

Not convinced? The Pimat document is full of diagrams, Kirlian photographs, charts and, of course, testimonials galore. Pains vanish, infertile women conceive babies, bedwetting disappears (surely of comfort to Pimat!) and people experience more lucid dreams because "pyramids generate fractal energy fields." The only warning is that the efficacy of Pimat may be reduced if the bed is against a wall that has a fridge or TV on the other side. In such a situation, you are urged to "contact your dowser or feng shui consultant." Needless to say, Pimat has no side effects. That, unquestionably, is a true statement.

While Pimat can perform numerous miracles, I was disappointed to learn that it may not be able to neutralize "geopathic stress." That, you should know, occurs when the earth's electromagnetic field becomes distorted. I learned that "the Earth resonates with an electromagnetic frequency of approximately 7.83 Hz (Schumann resonance), which falls within the range of (alpha) human brainwaves. Underground streams, sewers, water pipes, electricity, tunnels and underground railways, mineral formations and geological faults distort the natural resonance of the Earth thus creating geopathic stress." Pimat's powers are apparently tested to the extreme when dealing with geopathic stress, but luckily the "HELIOS 3 Geopathic Stress & EMF Home and Office Harmoniser," which can be plugged into any electrical socket, can ride to the rescue.

By this time, my head was spinning, perhaps from all the "radiesthetic energy" that I had absorbed from fondling Pimat. But I thought there was perhaps a teaching opportunity here. I suggested to the ladies that we run a little test. We would enlist a number of subjects who would sleep either on Pimat or on a piece of the same fabric without the red dots. Both the Pimat and the control fabric would be sewn into coded cotton envelopes so nobody would know who was sleeping on what. I had been assured by the manufacturer that the cotton envelopes would not interfere with the therapeutic effect. Indeed, given that the product is supposed to work its magic when placed under the bedsheet, they could hardly claim that the wondrous neoenergy could not pass through a cotton envelope.

I managed to enlist about eighty senior subjects who, before even being told what the experiment would be, were asked to fill out questionnaires about their health and energy status every morning for two weeks. For the following four weeks, half the subjects slept on Pimat and half on the placebo fabric. They then switched samples and carried on for another four weeks, filling out questionnaires every morning.

As I had expected, the presence of red dots made absolutely no difference. But there was a significant difference between the first two weeks, when there was no intervention of any sort, and the next eight weeks. Whether they slept on an "active" Pimat or on the placebo version, people claimed that they had slept better and woke up with more energy. Some even claimed that they had been able to perform conjugal activities during the night with greater vigour.

To their credit, the two women recognized that they had witnessed an example of the placebo effect and decided not to market Pimat, as had been their intention. Others may not be so insightful. One Pimat devotee proclaims that her alternative doctor confirmed that her immune system was now at 80 per cent. I wonder what breakthrough technology she uses to make such a measurement. In any case, she was so impressed that she ordered twenty-two Pimats for her patients at about $20 each. Of course, couples sleeping together need two Pimats.

Although our study of Pimat was not rigorous enough for publication, the fact that the importance of carrying out controlled trials had made such an impression on my new acquaintances did leave me with a degree of satisfaction. I was left with something else as well: a number of Pimats. And I have since discovered that they really do work. For cleaning computer screens.

What celebrity chef made attempts to improve the food served in American school cafeterias?

That would be Jamie Oliver. This British chef opened up quite a can of worms with his plan to improve the food served in Los Angeles schools. Actually, worms would probably be an improvement over some of the fatty processed food served, but after initially agreeing to the filming of Jamie's popular *Food Revolution* program inside its schools, the Los Angeles school board decided that, while the chef's ideas for improved nutrition would be welcome, his cameras would not be. Apparently, they had heard that a similar venture by Jamie in Huntington, West Virginia, left a bad taste in some administrators' mouths.

Let me state for the record that I like Jamie Oliver. I value his attempts to improve students' health through his "revolution." But I'm revolted by his "science"—or, I should say, his lack of it. No big surprise here, I guess, given that Jamie left school at the age of sixteen to pursue his culinary interests at Westminster Catering College, where I suspect chemistry was not the highlight of the curriculum.

Unlike many self-proclaimed nutritional gurus, Jamie is not an extremist. While he favours organic ingredients when possible, he doesn't espouse a vegan diet and is not averse to a hamburger. He just wants that hamburger to be made of proper ground beef, not various meat by-products. And if kids are to have an ice cream sundae, he'd rather "it didn't contain shellac, hair or beaver glands." It is with statements like this that Jamie muddies the waters.

While the Los Angeles school board was reticent about allowing him to exercise his culinary talents in the kitchen, one school agreed to let Jamie "teach" a science class about food. And these kids were sorely in need of some education along these lines, considering that some thought honey comes from bears and chocolate is pumped from a chocolate lake. Jamie's science class, though, came down to frightening students away from processed foods with a dramatic but nonsensical demonstration.

"Do you know what is in your ice cream sundae?" Jamie asked the class as he prepared to reveal the "truth." Out came a blender and in went a mix of live lac bugs, human hair and feathers. Surely beaver glands would have been thrown in had they been available. Instead, Jamie had to make do with a stuffed beaver looking on forlornly, as if deprived of his anal glands by nasty food chemists looking to improve the flavour of ice cream. What a blend of nonsense

Let's start with the shellac, a secretion of the lac bug that, in a purified form, can indeed be applied as a coating on the candy topping that decorates sundaes. But implying that chopped live insects are an ingredient in sundaes is ridiculous. As ridiculous as the theatrics with hair and feathers. Here the reference is to cookie dough that may be found in some ice creams and is often formulated with L-cysteine, an amino acid that improves texture. This compound can be readily isolated from the mix of amino acids produced by chemically breaking down proteins. Indeed, both hair and feathers are composed of proteins and can serve as the raw materials for the production of L-cysteine. These days, cysteine is actually made by a fermentation process, but its origin is really irrelevant. What matters is the final product. And L-cysteine is a harmless, approved food additive.

Any suggestion that duck feathers are added to ice cream is quackery.

On to the beaver glands. Jamie debuted this piece of puffery on *The Late Show*. Host David Letterman has been sworn at by Cher, instructed on the use of cucumbers in the bedroom by Dr. Ruth Westheimer and endured exposure to a variety of excreta from his animal guests. But rarely has he expressed the kind of shock we saw when Chef Oliver blurted out that cheap strawberry syrup and vanilla ice cream can contain beaver anal glands.

Audience reaction mirrored Letterman's, and the episode triggered predictable exchanges on the web expressing outrage about "what they're putting into our food." Some people wondered about just how many beavers sacrificed their lives to produce that pint of ice cream in the fridge.

The answer to that is: none. But it is true that when beavers are trapped for their pelts—admittedly not a pleasant thought—two small glands near the anus that produce a territorial marker called castoreum are removed, and their contents are extracted with alcohol for commercial use. A few parts per million of purified castoreum may be one of the ingredients included under "natural flavour" on a jam or ice cream label. But that's a long way from mixing beaver glands into ice cream.

So we have an interesting philosophical question here: Does the end justify the means? Is it acceptable to improve people's diet by evoking the yuck factor through false information? Jamie Oliver strives to feed people a proper diet, which is great, but he is also putting their brains on a diet devoid of science. In my view, teaching the wrong thing is never right.

I wonder what Jamie thinks about drinking the mammary gland extract of a cow? Or eating the ovum of a chicken? Has he ever had any escargots? It's also interesting to note that while he was terrifying Letterman with prospects of beaver glands in his ice cream, Jamie was cooking up mussels. Have you ever seen what raw mussels look like? Beaver glands appear positively charming next to this slimy tissue.

If you want to worry about something in ice cream, worry about the high sugar and fat content. As far as the beaver glands go, well, obsessing about them is, let us say, anal.

How did Tanzanian pastor Ambilikile Mwasapile rise to fame?

Mwasapile, known to his followers as Babu (old man), claims that a potion he brews from the *mugariga* plant (*Carissa spinarum*) cures all diseases, including AIDS, cancer and diabetes. How does a pastor with no scientific education come to make such a claim? It seems he got advice directly from God in a dream. That apparently is convincing enough for the desperate. They come in droves to the remote village of Samunge, where Mwasapile administers his cure, which he says only works when given by him personally. Furthermore, if someone jumps the queue while waiting for the magical cup of herbal brew, it won't work. And there certainly is a queue.

At one time in 2011, the lineup of vehicles carrying some 24,000 patients stretched for fifty-five kilometres. But driving over hundreds of kilometres of poor roads and waiting for days without access to clean water, toilets or shelter is apparently no deterrent when it comes to the allure of a cure-all. Unfortunately, over a hundred people waiting for that cure never managed to make it to the front of the line. They died. Mostly, they were seriously ill, taken out of hospitals by relatives who believed they were more likely to be cured by the pastor than by doctors.

At one point, Mwasapile appealed to the authorities to stop the traffic for a week so he could clear up the logjam. "This is a pathetic situation," he complained, "and something should be done to stem the crisis." He was right. It was a pathetic situation. But not because of the traffic jam. It was pathetic because sick people were being misled. And in some cases, it cost them their lives. Documented reports attest to AIDS patients refusing to take their antiviral drugs after being told by Babu that they had been healed of the disease.

So, people die waiting for treatment, people die because they think they have been cured by the treatment, and people pay outrageous amounts for transportation to Samunge. Not a pretty picture. But is it possible that it's all countered by this lowly, humble pastor having actually stumbled on a remedy that works? Herbal extracts can certainly have physiological effects. However, the claim that diseases can be cured by one single administration of an oral remedy immediately puts it into the implausible category. Medications just don't work like that. And, of course, different diseases with different causes are unlikely to be cured by the same remedy.

Carissa spinarum does have some biologically active ingredients. No great surprise here; virtually all plants have some. Studies show that extracts can help heal burns on mice, that they have free-radical scavenging activity and that they can produce an anticonvulsant effect when seizures are artificially induced in mice. But there is no evidence of healing any disease. And who knows how much of any potentially active ingredient there may be in a cup of Babu juice?

So, if a meaningful biological effect is unlikely, how come thousands and thousands flock to Mwasapile's humble "clinic"? Would they do so if there were no evidence that the treatment works? Of course they would! People flock to homeopaths, reflexologists, "distance healers" and "energy therapists" despite lack of evidence (that's right, there is no convincing evidence for homeopathy). They're seduced by a mix of anecdotes, conditions that improve by themselves, the placebo effect and faith in the treatment. Indeed, Pastor Mwasapile has said that only those who go to him in total faith can be cured. What we have here is time-honoured faith healing, with all its benefits and pitfalls.

What would you do with soursop?

You would eat it or juice it. Soursop is the fruit of a tropical tree, also known as custard apple, graviola or Brazilian pawpaw. It's a particular favourite in the West Indies and South America, both for its tangy taste and its supposed medicinal properties. If you can think of a condition, chances are someone will have reported that it can be treated with graviola. Either the fruit, the juice or teas made from the leaves of the tree on which the fruit grows have been anecdotally reported to help treat diarrhea, digestive problems, parasite infections, diabetes, asthma, colds, arthritis, high blood pressure, fever and anxiety.

Whenever I hear such a wide range of claims on behalf of one particular substance, my alarm bells start to chime. The ailments described have a variety of causes, and it is most unlikely they all respond to a single intervention. Our bodies just don't work like that. Asthma and diabetes, for example, are unrelated, and require different forms of treatment. Of course, with natural products such as graviola, there is always the argument that they are composed of hundreds of different compounds, and therefore it is possible that it contains some that may treat asthma and others that are effective for diabetes control. Possible, yes. Likely, no.

A more probable explanation is that the reported benefits are due to a blend of wishful thinking, the resolution of self-limiting conditions, unconfirmed anecdotes, and, of course, that good old standby, the placebo effect. Dozens of other plants, fruits and herbs that grow in the Caribbean and South America have histories as "cure-alls" as rich as that of graviola. But folklore is not evidence, no matter how compelling some individual testimonials may sound. Especially when it comes to serious diseases such as cancer. And yes, graviola is supposed to cure that as well.

Indeed, it is described as a cancer-killing dynamo. It attacks cancer safely and effectively without extreme nausea, weight loss or hair loss. It protects your immune system so you avoid deadly infections. You feel stronger and healthier throughout the course of the treatment. It effectively targets and kills malignant cells in twelve types of cancer, including colon, breast, prostate, lung and pancreatic cancer. It does not harm healthy cells! It is up to ten thousand times stronger in slowing the growth of cancer cells than Adriamycin, a commonly used chemotherapeutic drug. And, it is all natural!

Doesn't that sound fantastic? So how come you haven't heard of this miracle? The "spine-chilling" answer to that question, according to the promoters of graviola supplements, "illustrates just how easily our health—and for many, our very lives—are controlled by money and power." Big Pharma, they say, "is doing everything in its power to keep this natural astonishing cancer cure under wraps in order to protect the enormous profits it reaps from its toxic chemotherapy drugs that do little more than poison patients."

Aha! So, *that's* why you haven't heard about it. But how come I have? I guess I'm one of the lucky ones who received a brochure from the "Health Sciences Institute," featuring a captivating article entitled "Beyond Chemotherapy: New Cancer Killers, Safe as Mother's Milk." I was also informed that I could view a video about this stunning breakthrough, but I had better do it quickly, because it is unclear how long they will be able to prevent Big Pharma from shutting it down. Furthermore, the Health Sciences Institute has also been able to secure a "limited supply of graviola extract grown and harvested by indigenous people in Brazil."

Needless to say, I was intrigued. How did I miss this? Why wasn't this breakthrough trumpeted in medical journals around the world? Is Big Pharma really that effective at silencing the cutting-edge researchers who discovered the miraculous properties of this natural cancer treatment? Well, it turns out that the Health Sciences Institute isn't any kind of institute, and it isn't very scientific. It's a sales outfit.

As is often the case with such "wonder products," unethical promoters dredge the scientific literature to find some little grains of truth that they can then ferment into a seductive potion. And there is nothing more seductive, or more marketable, than a "secret cancer potion." In this case, those grains of truth were found in some research carried out at Purdue University back in 1997. A number of compounds isolated from soursop, called annonaceous acetogenins, were tested for their ability to kill cancer cells in the laboratory. The focus was on a particular type of cancer cell that was resistant to the effects of such common chemotherapy drugs as Adriamycin. Such resistance is not common, but nevertheless it is of academic interest. One of the compounds in graviola, bullatacin, was eventually identified as being effective in killing the resistant cancer cells.

Experiments like this are performed around the world regularly, and thousands of compounds with cancer cell–killing activities have been discovered. But they rarely progress to anything substantive in terms of human treatment. All that such preliminary findings mean is that further studies may be warranted to investigate whether there is any effect in animals. If that can be documented, then human trials may be indicated. But as far as graviola is concerned, nothing further has been published. Never mind human trials, there aren't even any *animal* trials that have been published. And, of course, we don't know whether graviola, when used as a drug, is free of side effects. There have actually been some reports of Parkinson-like symptoms after taking certain graviola extracts.

None of these concerns have prevented the energetic marketing of various graviola products with headlines such as "Deadly Conspiracy Exposed." The conspiracy is supposed to involve pharmaceutical companies that want to keep us from finding out about the graviola miracle. I think the only conspiracy here that needs to be exposed is the one that unethical marketers engage in when they try to capitalize on the desperation of cancer victims. Soursop may be a delicious fruit, but claims about its ability to cure cancer leave a sour taste in the mouth.

In 1987, Leonid Tenkaev, a Russian factory worker, claimed to have become a "human magnet," able to stick metal objects to his body. What event, according to him, caused the development of this unusual physical attribute?

Leonid Tenkaev and his wife claim that the nuclear accident at Chernobyl, in the Ukraine, transformed them into human magnets. Pictures of the couple show spoons, keys and even irons sticking to their bodies as if they were glued there. Perhaps they were. From the photos, one cannot tell. The Tenkaevs, however, are not the only ones to claim to have such magnetic powers. Indeed, the web is ablaze with pictures and videos of "human magnets" plastered with everything from coins to cell phones. But for these living curiosities, it is cutlery that seems to have a particular appeal. However, human magnets, unlike metal-bending psychics, don't disfigure these utensils; they just attract them. Explanations abound as to how bodies can generate electromagnetic fields, with some lucky individuals apparently being blessed with particularly potent ones. Spoons probably have nightmares about psychics and human magnets and their "energy fields." Scientists, too.

The human body does generate tiny fields, but these are way too weak to attract metals. Curiously, human magnets also claim to attract plastic objects, which have no magnetic properties. A tough one to swallow. One facet of human magnets is, however, evidence-based. They certainly attract attention. Mostly from gullible reporters who do not realize that there are various ways of performing seemingly remarkable feats without having to invoke paranormal explanations. All that is required is familiarity with the principles of deception—something that, of course, is in the domain of conjurers. Could that be why no human magnet has been able to claim the million dollars offered by the James Randi Educational Foundation for any demonstration of a paranormal phenomenon under controlled conditions?

Perhaps the most famous human magnet is Miroslaw Magola, a Pole who now lives in Germany but travels the world searching for "universal truths." It seems that in 1992 he discovered an ability to counter the laws of nature. Magola can cause objects to stick to his body simply through "thought power." Judging by the pictures he wildly distributes on the web, Magola specializes in metal bowls with smooth surfaces. He works himself up into some sort of magnetic mental frenzy and then places his palms on the bottom of upside-down bowls and picks them up, apparently defying gravity. He can also plant the bowls on his forehead—ready for use if there's a leak.

Magola appears to be reproducing an effect that has been performed by entertainers and fake psychics (surely an oxymoron) for over a hundred years. It does make for a great spectacle. Another human magnet, apparently on the cutting edge of science, proudly shows off a couple of meat cleavers stuck to his bare chest. Of course, it is not the edge but the flat side of the cleaver that is stuck to the skin. And therein lies a clue: the objects that stick to these human magnets always have a smooth surface.

Have you ever cut a potato with a sharp knife that has a wide blade? It can be quite a challenge to unstick the knife from the potato. Smooth surfaces brought together will stick, especially when they are separated by a thin layer of liquid. Pressing smooth objects to greasy skin creates a suction cup–like effect, especially when the subjects tilt themselves backwards, as they tend to do. And some people really do produce especially sticky sweat. There's even a condition known as acquired cutaneous adhesion syndrome.

Isn't it interesting that the human magnets are always hairless? And why is it that the magnetic attraction cannot pass through fabric? Authentic magnets don't have that problem. James Randi has debunked human magnets numerous times in lectures and on television simply by asking the claimants to dust themselves with talcum powder. They always fail to perform. But Magola claims that he has debunked the debunkers by applying talcum powder to his hands, and he even circulates a video as proof.

The video is totally unconvincing. He applies talcum powder to his right hand, wipes off most of it, and then proceeds to pick up the upside-down bowl with his left hand, which appears to be strangely curled as if to hide something. Perhaps a thin layer of grease? And why don't human magnets attract odd-shaped objects, or balls? Could it be because surface adherence is very much dependent on the size of the contact surface? And how come a compass needle doesn't point towards these magnetic people?

Magola claims that he is ready to meet the James Randi Educational Foundation's challenge and collect the million-dollar prize for successful demonstration of a paranormal phenomenon. The truth is that he has not applied for the challenge even after persistent requests to do so. Could it be that, under test conditions, the power would disappear? I suspect so. Especially with conjurers around who are familiar with magician's wax and various other methods of creating the illusion of defying gravity. After all, levitations and suspersions are the bread and butter of magicians.

Of course, magicians do not claim to be able to suspend the laws of nature; they just claim to provide wholesome entertainment using effects well within the boundaries of science. Unlike the human magnets, who sully people's minds by pretending to have paranormal powers. I find their claims of attraction repulsive.

How do you convert lime water into milk of lime?

There are no fruits or fruit juices involved in this question. "Lime water" refers to a clear solution of calcium hydroxide in water. Passing carbon dioxide gas into this solution results in the formation of insoluble calcium carbonate. It is this precipitate of calcium carbonate that gives the appearance of milk, hence the term *milk of lime*. Breath contains carbon dioxide, so blowing into lime water through a straw causes the solution to turn cloudy. This is an instructive chemistry experiment that demonstrates the use of a simple reaction to test for carbon dioxide in the breath. It also leads naturally into a discussion of solubility. In the wrong hands it can also lead to an ingenious scam.

Back in the heyday of the travelling medicine shows, lime water was sometimes used as a clever marketing tool. Medicine shows crisscrossed the country in the fashion of a circus, and indeed had a circus element to them. Crowds were attracted by singers, comedians and various novelty acts, but those were just a buildup to the finale, the sales pitch from the "Professor." It was his job to convince the audience that the patent medicine he was hawking was the answer to all their ailments. There were the usual planted shills who would enthusiastically describe how they rose from their deathbed after consuming the Professor's snake oil. And there would be some convincing demonstrations, one of which involved a witty use of lime water.

An audience member was invited up on stage and asked to blow into a solution that "detected disease." Sure enough, when the poor chap exhaled into the lime water, it turned cloudy. The Professor would then announce that an ailment had been discovered, but luckily a remedy was at hand. He proceeded to pick up his bottle of medicine, added a few drops of it to the cloudy solution, and as if by magic, the solution turned clear. The message was also clear: taking the medicine would have the same effect on the poor victim's insides.

A pretty convincing demonstration to people who, of course, had no idea of the chemistry involved. The miraculous medicine was nothing other than a dilute solution of acetic acid, commonly known as vinegar. Acetic acid reacts with calcium carbonate to form water-soluble calcium acetate and carbon dioxide. The bubbling produced by the carbon dioxide likely added to the perception that something truly therapeutic was happening.

While there was no therapy, there was some interesting chemistry going on. Basically, it is the same reaction as occurs when vinegar is used to remove the scale deposits from inside a kettle. Those deposits consist of calcium carbonate, which forms when heat causes soluble calcium bicarbonate, naturally present in water, to decompose into insoluble calcium carbonate. Vinegar can dissolve the calcium carbonate and clean the kettle. It also cleaned out the poor sucker's pockets, but not his insides.

Where would a travelling medicine show pitchman have gotten lime water? It's simple enough to make. All you need is some calcium oxide, or quicklime, which will readily react with water to form calcium hydroxide. And where did the calcium oxide come from? That compound, a major component of cement, has been produced since ancient times by the heating of limestone. Making lime water requires some care, because the reaction of quicklime with water is extremely exothermic. In fact, the reaction can produce enough heat to cook food. That's why quicklime has been used as a convenient portable source of heat—for example, in self-heating cans.

It also lends itself to a neat chemistry demonstration. Place some calcium oxide on a plate, add water, and then break an egg over the mixture. It will fry. I bet some quack could put that reaction to an ingenious and undoubtedly nefarious use.

TO YOUR
HEALTH

Our body uses the family of cytochrome p450 enzymes to deal with toxins. What does the number 450 refer to?

The *p* in the name of this family of enzymes refers to *pigment,* and the number 450 refers to the specific wavelength of light used to spectroscopically identify these coloured (*chrome*) molecules that are found inside cells (*cyto-*). A spectroscopic study involves exposing a sample to various wavelengths of light and determining which wavelengths are absorbed, a common laboratory identification technique.

We are constantly exposed to foreign substances that have to be dealt with before they can cause harm. Aromatic hydrocarbons in smoke, bacterial toxins, remnants of industrial chemicals, as well as food components such as caffeine, tannins and alkaloids have to be eliminated before they engage in harmful reactions. Drugs are also viewed by the body as intruders that need to be eliminated.

Many of these substances that we collectively call *toxins* are relatively insoluble in water, and therefore present a challenge for elimination through the kidneys. The cytochrome enzymes add oxygen atoms to these molecules, increasing the ease with which they can be flushed out of the system. In some cases, the addition of oxygen just makes the compound more soluble, and therefore more readily excreted through the urine. In others, the oxygen atom introduces a reactive site that molecules such as glucuronic acid can latch onto. Since glucuronic acid is highly soluble, it acts as sort of a ferry to help eliminate compounds that it has been able to grab. Cytochrome enzymes are found mostly in the liver, the organ charged with intercepting dangerous substances before they can enter the circulation and wreak havoc with more susceptible tissues.

But these enzymes can also be looked upon as the proverbial double-edged sword. A drug, the painkiller acetaminophen for example, is a candidate for detoxication as it passes through the liver. The reason such medications have to be taken every few hours is that they are constantly being removed from circulation by the cytochrome enzymes. Indeed, the reason people have different sensitivities to drugs is a consequence of having a different profile of cytochrome p450 enzymes.

Another problem is that the oxygenated form of a compound may sometimes be more toxic than the original. For example, benzopyrene in smoke, a relatively harmless intruder, is converted by cytochrome enzymes into a form that can react with and disrupt DNA. In this case, the oxygen provides a "handle" that DNA can seize.

Interaction of toxins with cytochrome p450 can raise yet another concern. Furanocoumarins, naturally occurring compounds in grapefruit juice, can react with these enzymes and prevent them from carrying out other detoxicating tasks. That's why patients are told not to take their medications with grapefruit juice. Since the enzymes are not available to eliminate the drug, it can build up and have a greater effect than intended. Certain blood pressure–lowering drugs—felodipine, for example—can then have the effect of decreasing the blood pressure to dangerous levels.

Finally, there is also the possibility that some substances the body perceives as toxins can induce the formation of higher-than-normal levels of cytochrome enzymes in an attempt to eliminate the intruder. In the process, other substances may be undesirably eliminated as well. There have been cases of organ rejection attributed to patients taking antidepressants such as St. John's wort when they were being treated with the immunosuppressive drug cyclosporine. The increased levels of cytochrome enzymes can cause more rapid elimination of cyclosporine, resulting in organ rejection.

Putrescine and cadaverine are nasty-smelling compounds that are components of bad breath. How are they produced in the mouth?

Bacteria that frolic in the absence of oxygen, known as *anaerobic bacteria,* generate putrescine and cadaverine as they break down some of the proteins present in food. Probably to the surprise of many, our mouth teems with some six hundred different species of bacteria. Every millilitre of saliva harbours over 100 million of the tiny microbes, which are mostly harmless, but can cause problems under certain conditions. Tooth decay and gum disease are the most common examples of such problems, but bacteria infiltrating the bloodstream through lesions in the mouth have also been associated with heart disease.

Oral bacteria can attach themselves to the smooth surface of teeth and form a layer that then traps various proteins, fats and carbohydrates to form a film called plaque. At first, this *biofilm* is soft and easily removed by brushing, but if not removed within forty-eight hours, it hardens, eventually forming a tough, difficult-to-remove substance called tartar. The buildup of plaque creates an oxygen-deficient environment next to the tooth surface, ideal for the growth of anaerobic bacteria. These then begin to digest the proteins around them, breaking them down to their component amino acids, which are then further metabolized to the delightful putrescine and cadaverine. Ornithine, for example, yields putrescine, and lysine is the source of cadaverine.

As the names imply, these are unpleasant smelling compounds, with putrescine being responsible for the smell of rotting meat and cadaverine contributing to the fragrance of decaying bodies. Not exactly what we want emanating from our mouths. Add to these such other malodorous bacterial byproducts as dimethyl sulphide, isovaleric acid, hydrogen sulphide and skatole, and we have a potent blend that can make for frightful halitosis. Dimethyl sulphide is the aroma of boiling cabbage, isovaleric acid is the odour of sweaty feet, hydrogen sulphide that of rotten eggs, and skatole is one of the components of fecal aroma.

Obviously, we would like to limit the bacterial action that can produce these disturbing odours, which is why we want to keep our mouths free of plaque with proper frequent brushing.

What smell comes from a solution of salt in water that has sat for a few hours?

The telltale odour of iodine. Potassium iodide is commonly added to salt as a protection against goitre. But iodide reacts slowly with oxygen to form iodine, hence the appearance of the smell. Before going into the interesting chemistry employed in trying to keep salt from losing its iodide content, a little background comes in handy.

The thyroid gland located in our neck produces hormones that control the rate of energy production in all cells in our body, and therefore influences the functioning of all organs. Insufficient production of thyroid hormones can cause weight gain, lethargy, constipation, cold clammy skin and perhaps even hair loss and premature greying. In utero it can cause cretinism. There are various reasons for hypothyroidism. The most common form is known as Hashimoto's disease and is caused by chronic inflammation of the thyroid gland. It is treated by giving the patient thyroid hormone.

But there is another possible cause for hypothyroidism: lack of iodine in the diet. The thyroid hormones have iodine in their molecular structure, and if there is not enough iodine in the diet, the thyroid gland cannot make enough hormone. It struggles to withdraw iodine from the blood, and in the process enlarges, producing goitre, a typical sign of lack of iodine in the diet. When we use the term *iodine* in this context, it is incorrect. In the soil, iodine occurs in its ionic form, known as iodide. That's the form that is added to salt. But it is common practice to use the term *iodine* in reference to what is found in the soil, although the correct term should be iodide.

Until the early part of this century, the American Midwest was known as the Goiter Belt because of the lack of iodine in the soil. Crops, and the animals that fed on them, therefore had little iodine. Scientists wondered why goitre was endemic in this area and was almost never found near coastal areas. By 1920 it had become evident that the connection involved iodine. This also explained why goitre was seen in areas where cabbage was a staple in the diet. This vegetable contains isocyanates that interfere with the absorption of iodine. "Cabbage goitre" is usually found in areas where the soil is already deficient in iodine.

When people found out about this, they decided to take matters into their own hands. Well, not exactly their hands: people hung bottles of iodine around their necks. Some even started to eat iodine to the extent that they produced a *hyper*thyroid condition. And then, in 1924, Michigan began to experiment with adding sodium iodide to salt, and Rochester, New York, added it to drinking water. The solution to the goitre problem was under way.

Today, salt with added potassium iodide is common. But there is a problem. Iodide is slowly converted to iodine in moist air. Since iodine is volatile, salt slowly would lose its power to protect against goitre. It is therefore common practice to add iodine stabilizers to iodized salt. These are substances that convert iodine back to iodide.

Sodium thiosulphate used to be added, but it sounded too chemical and people became worried. Manufacturers switched to dextrose, which is as effective and sounds more innocuous. Sodium bicarbonate is also added, because the oxidation of iodide occurs readily in an acid solution and not in a base. The bicarbonate produces basic conditions. Sometimes disodium phosphate or sodium pyrophosphate are used to provide the alkaline conditions. These are also sequestering agents, which bind trace metals that catalyze the oxidation of iodide to iodine. Not all salt has stabilizers, and these are the ones most likely to give rise to an iodine scent.

Unfortunately, iodine is not added to salt everywhere. In India, an estimated 250 million people suffer from iodine deficiency, which means decreased motor skills, low IQ and poor energy levels. About nine million children are born as cretins. Because of India's humid air, salt is usually sold in large crystals, which are less likely to absorb moisture. These are sometimes sprayed with potassium iodate, but that makes the crystals look dirty, and users wash them before crushing them. Not a smart thing to do.

✦

What supposed aphrodisiac is used by dermatologists to treat warts?

Cantharidin, a compound extracted from blister beetles, has a history of folkloric use as an aphrodisiac, a substance that supposedly increases sexual desire. Not only does it not do that, it may eliminate all desire permanently. When ingested, cantharidin can kill. But when used in small doses topically, it can be an effective treatment for warts.

Blister beetle extract is sometimes referred to as *Spanish fly,* which is curious, since the beetles are neither flies, nor are they Spanish. Some 1,500 species of cantharidin-producing beetles can be found around the world. They do not bite or sting, but can cause a great deal of physiological mischief if they are accidentally—or in some cases, deliberately—ingested.

Cantharidin is synthesized by the male beetle and is given as a nuptial gift to the female during copulation, for the purpose of protecting her eggs. If a predator bites into a blister beetle egg, it learns to never do it again—cantharidin causes terribly painful blisters. But that is not all it does. In humans, it can cause a burning sensation of the urinary tract that, in men, can provoke an erection. This effect is responsible for cantharidin's undeserved historical reputation as an aphrodisiac that dates back all the way to ancient Roman times.

Augustus Caesar's wife, Livia, supposedly slipped ground beetles into food to entice men into indiscretions for which they could be blackmailed. There are also accounts of Louis XIV's lust for women being aided by cantharidin, but the most famous proponent of Spanish fly as a sexual stimulant was the notorious Marquis de Sade, who is said to have given prostitutes cantharidin-laced pastilles to "set them on fire." It set them on fire, all right—with burning abdominal pain. The marquis and his valet were accused of poisoning the women, and they quickly fled to Italy. They were sentenced to death in absentia.

Cantharidin also resulted in a death sentence for an American fisherman, but in a different fashion. In 1954, for some bizarre reason, the man came to believe that blister beetle extract would attract fish. He shook some ground-up dried beetles with water in a bottle, using his thumb as a stopper. Immediately after, in the process of handling his hooks, the fisherman pricked his thumb, and as most would do, proceeded to suck on it. There was enough residual cantharidin on the thumb to kill him. It doesn't take much—the fatal dose of cantharidin is in the range of 10 to 65 milligrams.

According to some accounts, blister beetle extract was one of the components in Giulia Tofana's infamous "widow-maker" poison in the seventeenth century, although most evidence points to the poison actually being a mixture of arsenic, lead and belladonna. Whether Spanish fly was a component of Aqua Tofana is contentious, but there certainly was a belief that cantharidin was being used as a poison. Indeed, attempts were made to try to identify it in the bodies of victims who were thought to have been poisoned. A piece of an internal organ from the deceased would be extracted with oil and the solution placed on the shaved skin of a rabbit to see if blistering would be produced.

Pure cantharidin was first isolated in 1810 by the French chemist Pierre-Jean Robiquet, who identified it as the substance responsible for the blistering properties of the beetle eggs. The exact molecular structure was not proposed until the mid-twentieth century, and was shown to be correct when American chemist Gilbert Stork synthesized cantharidin from simple molecules. This was only of academic interest because there is no need to synthesize cantharidin commercially. Whatever amounts are needed for dermatological use or for animal husbandry can be extracted from beetles, which are not hard to find. Why in animal husbandry? In some cases, mating needs to be encouraged. As in humans, cantharidin can produce an erection in males, prompting the animal to try to resolve the situation through mating.

Because of the myth of the aphrodisiac effect, various varieties of Spanish fly are available for purchase online. It is unlikely they actually contain any cantharidin—which, of course, is a good thing, given the danger of the actual compound. Mostly, they are herbal concoctions based on cayenne-pepper extract that will produce a mild burning sensation, mostly in the wallet. There are also some Spanish fly candies available for purchase. Very ordinary candies wrapped in paper adorned with the name Spanish Fly.

In Germany, homeopathic remedies based on cantharidin are available for the treatment of urinary tract infections. They cannot cause any harm because homeopathic remedies are diluted to an extent where either none, or a trivial amount, of the original substance remains. Needless to say, such products cannot produce a physiological reaction and rely totally on the placebo effect.

A 0.7 per cent solution of cantharidin applied to a wart, however, *can* produce a physiological effect. The compound can cause skin cells to disengage from each other, resulting in the breakdown of a wart. Application must be done by a physician, because the amount applied and the time the solution is left on the skin are critical.

The prototype for what family of drugs was isolated in 1973 from *Penicillium citrinum*, a mould found on a Japanese orange?

Mevastatin, isolated from a mould, was the first of a family of drugs that would come to be known as the *statins*. It's a good bet that these drugs have played a significant role in the 50 per cent decrease in deaths from heart attacks and strokes we've seen since the 1980s. Their discovery makes for an interesting story.

By the 1970s, the ability of moulds to produce biologically active compounds was well established. Alexander Fleming's 1928 discovery of a mould that produced a substance effective against bacteria had already borne fruit with the widespread introduction of penicillin. But unlike Fleming's discovery, that of mevastatin by Dr. Akira Endo was not accidental. Endo, working for the Japanese pharmaceutical company Sankyo, was actively searching for a drug to lower cholesterol levels in the blood.

As Dr. Endo tells the story, his interest in moulds was sparked by walks in the woods with his grandfather, who pointed out a mushroom that apparently killed flies, but not people. Later, after graduating from Tohoku University, Endo's appetite for mould research was further whetted by reading about Fleming's work. And the research paid off when, after investigating some 250 kinds of fungi, Endo found one that produced an enzyme capable of breaking down pectin, a carbohydrate found in fruits. This had a commercial application, since it could be used to make fruit juice less pulpy.

As a reward, Sankyo sent Dr. Endo to the Albert Einstein College of Medicine in New York to further his biochemical expertise. It was here that he got involved with cholesterol research, already a hot topic in America, where the link between heart disease and elevated levels of blood cholesterol had received a great deal of publicity.

At the time, it was already known that cholesterol, an essential biochemical, was synthesized in the liver. It was an integral part of cell membranes and served as the starting material for the body's synthesis of a variety of compounds ranging from cortisol to the sex hormones. But excess cholesterol in the blood could deposit in the coronary arteries, impairing the flow of blood and increasing the risk of a heart attack. One way around this problem would be to interfere with the activity of a key enzyme involved in the liver's synthesis of the compound.

Endo knew that bacteria, like humans, are capable of making the cholesterol needed for the integrity of their cell walls. His thoughts turned to fungi, reasoning that since these are sometimes attacked by bacteria, they may have evolved ways to protect themselves from the invaders. He wondered if some fungus might have evolved a way to fight bacteria by producing a substance that would impair the bacteria's synthesis of the needed cholesterol.

In 1973, after testing more than six thousand fungal preparations, Endo finally found what he was searching for in a mould growing on a Japanese orange. From *Penicillium citrinum,* he isolated a compound capable of blocking the activity of a cholesterol-synthesizing enzyme found in pulverized rat livers. That enzyme goes by the tongue-twisting name of beta-hydroxy-beta-methylglutaryl-CoA reductase, or HMG Co-A reductase. Christened *mevastatin,* Endo's compound was tested in rats for its cholesterol-lowering effect. The results were disappointing.

But then fate intervened. A colleague had some hens that were about to be destroyed and offered them to Endo. In these birds, mevastatin worked spectacularly. Sankyo patented the drug, but was not particularly interested in pursuing this research for fear of side effects. After all, cholesterol was an essential biochemical, and preventing its formation might result in unwelcome consequences. Indeed, when given to dogs in large doses, mevastatin caused gastrointestinal lesions.

But Endo was convinced he was onto something, and he embarked on a secret experiment that would be inconceivable today. He contacted a physician who was treating patients suffering from extremely high cholesterol because of a genetic defect and convinced him to try mevastatin. It worked, reducing cholesterol by 27 per cent in nine patients! In the meantime, after having learned of Endo's work, researchers at Merck began a screening program to come up with a mevastatin-like drug that they could patent. In 1978, they hit pay dirt. From the *Aspergillus terreus* fungus, Merck chemists isolated lovastatin, approved in 1987 by the Food and Drug Administration as a cholesterol-lowering drug.

After extensive human testing, it hit the market as Mevacor. The product insert had to stipulate that, while the drug had been shown to reduce cholesterol, no clinical outcome could be inferred, since at the time no study had shown that lowering cholesterol reduced the risk of heart attacks. But that would come in 1994, when a landmark clinical trial using simvastatin (Zocor), another Merck drug, showed a significant protection against a second heart attack in patients who had already been afflicted once. Everyone now wanted to jump on the statin bandwagon.

Sankyo got back into statin research and came up with pravastatin (Pravachol), and Pfizer introduced atorvastatin (Lipitor), destined to become the world's best-selling drug, raking in some $12 billion a year. Some fourteen clinical trials, involving over 90,000 patients with cardiovascular disease, have now documented a 30 per cent reduction in heart attacks after treatment with statins. And it is difficult to say how many heart attacks may have been prevented by treating people who have high cholesterol but no existing cardio-vascular disease with statins.

Of course, statins do not come with a no-risk guarantee. Side effects can include insomnia, rashes and gastrointestinal problems, which are relatively minor. But there can also be serious muscle inflammation, hepatitis and kidney failure. As with any drug, the appropriate use of statins comes down to a proper risk–benefit analysis. It is best to reduce cholesterol with diet and exercise, which with dedication can be done. In fact, Dr. Akira Endo demonstrated that himself. Diagnosed with high cholesterol at the age of seventy-one, he undertook a program of exercise and lowered his LDL, or "bad cholesterol," level from 4.0 millimoles/litre to 3.4. When asked why he didn't go on a statin regimen,

Endo resorted to a Japanese proverb: "The indigo dyer wears white trousers."

Why did the motorist put pennies in his mouth when he saw flashing lights in the rear mirror?

Seeing the flashing lights of a police cruiser in the rearview mirror always revs up the pitter-patter of a driver's heart, especially when he has had one too many. In this case, the approaching police car prompted the driver's use of his mouth as a piggy bank. Fearing he would be accused of driving under the influence, he recalled something about pennies in the mouth fooling the Breathalyzer. Indeed, this bizarre idea has been floating around the Internet for a while. Where does it come from? Probably some student who didn't pay proper attention in organic chemistry class.

The term *alcohol* actually refers to a whole family of compounds that have a common structural feature: a carbon atom joined to an oxygen atom joined to a hydrogen atom. Ethanol, the alcohol we drink, is the most famous member of the family and has the formula CH_3CH_2OH. Alcohols undergo a variety of chemical reactions, including one with oxygen, appropriately called *oxidation*. In the case of ethanol, this results in its conversion to acetaldehyde. The reaction, however, does not happen spontaneously. It requires a catalyst.

In the body, that catalyst is an enzyme known as alcohol dehydrogenase. Once ethanol has been converted to acetaldehyde, a second enzyme, aldehyde dehydrogenase, goes to work and converts the acetaldehyde to acetic acid, which is either excreted in the urine or is further processed into carbon dioxide and water. As these reactions proceed, the alcohol concentration in the blood drops, and the effects of alcohol abate. This sequence of reactions is what is referred to as the metabolism of alcohol.

Ethanol can also be converted to acetaldehyde in the laboratory, with various catalysts being available. One of these is metallic copper. Just dip a penny in alcohol, hold it in a flame and the alcohol is readily oxidized to acetaldehyde. Undoubtedly, this is the reaction that triggered the copper-in-the-mouth myth. And a myth it is. The idea, of course, is that the copper pennies will catalyze the conversion of ethanol to acetaldehyde, leaving no ethanol left for the Breathalyzer to detect. But the major point that the organic chemistry student who formulated the copper advice failed to remember was that the copper had to be heated to a high temperature. So unless our inebriated driver was willing to put *red-hot* pennies in his mouth, there would be no fooling the Breathalyzer.

Neither would any of the other methods that supposedly eliminate alcohol from breath work. You can suck on breath mints, onions or batteries, but they will not alter the Breathalyzer reading. And if it reads over 0.08, meaning that you have more than 8 grams of alcohol per 100 millilitres of blood, you are impaired. Unless you've been hyperventilating just before a breath sample was taken—breathing pattern can actually influence breath testing for alcohol. Holding your breath for about thirty seconds before giving a sample can increase the reading by 16 per cent or so, while hyperventilating for twenty seconds before sampling can result in a reading that is up to 30 per cent lower than it should be. Officers, however, are trained to recognize such attempts at subterfuge. Running up a flight of stairs can also decrease readings by some 15 per cent, although this is hardly a technique that can be employed after being stopped by the police.

There are several different types of Breathalyzer on the market. The most common one is based on the oxidation of ethanol, this time using a platinum catalyst. Breath is exhaled into an electro-chemical cell equipped with two platinum electrodes—one positively, the other negatively charged. At the positive electrode (the anode), ethanol is oxidized to acetic acid. This reaction liberates electrons that then travel to the negative electrode (the cathode), where they are used in a reaction that converts oxygen to water. The electrical current produced by the electrons flowing from the anode to the cathode can be measured, and is representative of the amount of ethanol being converted to acetic acid. Since the ratio of alcohol in the breath to that in the blood is well known, the instrument can be calibrated to give a direct reading for alcohol in the blood.

Another type of Breathalyzer measures the absorption of infrared light by alcohol. Atoms held together by a chemical bond are always moving relative to each other. A spring with weights at the end would be an analogy. And specific bonds have specific vibrational frequencies. When a bond is exposed to electromagnetic radiation of a frequency that corresponds to its vibrational frequency, it will absorb some of the energy of the radiation. The vibrational frequencies of chemical bonds happen to correspond to frequencies that characterize the infrared portion of the electro-magnetic spectrum. Infrared energy is what we sense as heat.

In an infrared Breathalyzer, infrared light is passed through the sample, and the extent to which it is absorbed at the specific absorption frequency of the oxygen–hydrogen bond is measured. This can then be related to the amount of alcohol in the sample. It sounds very complicated, and it is. A further complication is that both types of Breathalyzer can give false readings when certain other compounds are present in the breath. Acetone, possibly present in the breath of diabetics, can interfere, as can alcohol from mouthwash or alcohol that has been burped up from the stomach. Because of these possibilities, a 15 per cent tolerance in readings is usually allowed for by the police. Many suggest that the 0.08 per cent limit is too high and that a more realistic maximum blood level should be 0.05. Vested interests argue against this, but the fact is that some 1,500 people in Canada every year are killed in alcohol-related car accidents.

Whether the limit of alcohol in the breath should be reduced is debatable, but what is not debatable is the ingenuity of the chemistry that has been developed to measure breath alcohol levels. And by getting inebriated people off the road, that chemical ingenuity saves lives.

Which vitamin can only be produced by bacteria?

Bacteria get a lot of bad press because they can cause a lot of nasty diseases. But they can also do good things. Like make vitamin B_{12}, without which we would perish.

The definition of a vitamin is a substance that is needed for our health, but which the body cannot make and which therefore must be furnished by food. Vitamins play different roles— sometimes multiple ones. Such is the case for vitamin B_{12}, which in terms of molecular structure is the most complicated vitamin. Indeed, it has such a complex structure that the total chemical synthesis of the vitamin reported in 1972 by Robert Woodward and Albert Eschenmoser is regarded as one of the epic feats of organic chemistry. The synthesis has no practical application, since vitamin B_{12} can be readily made by bacterial fermentation methods, but the reactions that were developed as a result of the ingenious and involved synthesis contributed greatly to the advancement of organic chemistry.

Without B_{12}, the brain and nervous system cannot function properly, and cells cannot produce adequate energy or synthesize

DNA. We also need B_{12} to make red blood cells. Without adequate numbers of properly functioning red blood cells, oxygen cannot be supplied to all the cells in the body that need it. We then have a condition known as anemia, which basically means oxygen deficiency. The type of anemia associated with B_{12} deficiency is known as *pernicious anemia* and occurs when the vitamin cannot be absorbed from the digestive tract into the bloodstream in adequate amounts.

The absorption of vitamin B_{12} is a complex process. The vitamin in food is bound to proteins that have to be broken down before it can be absorbed. This requires adequate acid production in the stomach, which can be a problem as we age, since acid production tends to decrease as we get older. But release of vitamin B_{12} from its protein coat is only the first step in the utilization of the vitamin. Absorption actually takes place in the small intestine, but it requires the presence of a special protein, called *intrinsic factor,* produced by cells in the stomach. Only when adequate amounts of both B_{12} and intrinsic factor are present in the intestine can absorption into the bloodstream take place. Anemia, then, can occur either when there is a lack of B_{12} in the diet, when there is reduced stomach acidity, or when the *parietal* cells in the stomach do not manufacture enough intrinsic factor. The latter can happen when, for some unknown reason, the body's immune system attacks and destroys the parietal cells. When anemia is caused by the loss of gastric parietal cells and the subsequent inability to absorb vitamin B_{12}, it is known as *pernicious anemia.*

Signs of vitamin B_{12} deficiency can take a long time to show up and can be subtle. For example, B_{12} is needed for the maintenance of myelin, the protective coating around nerve cells. If this wears away, tingling sensations, numbness and a burning feeling in limbs can occur, along with memory problems and mood disorders. The impairment of red blood cells' ability to carry oxygen is a more dramatic effect of vitamin B_{12} deficiency. In the absence of B_{12}, these cells grow to an abnormally large size as they try to capture more oxygen. The technical term used to describe this condition is macrocytic anemia.

Monthly B_{12} injections or high-dose oral supplements (0.5 milligrams) can be prescribed to correct a diagnosed deficiency, but many researchers feel that it is a good idea for everyone over fifty to take a B_{12} supplement that includes at least the daily value of 2.4 micrograms, although some experts recommend higher doses, up to 25 micrograms. The usual form of the supplement is cyanocobalamin, which interestingly does not occur in nature, but in the liver can be converted to methylcobalamin or adenosylcobalamin, which are the physiologically active forms. Several years' worth of vitamin B_{12} can be stored in the liver, so it may take a long time for a deficiency to occur.

Actually, the term *vitamin B_{12}* can refer to any one of a class of chemically-related compounds that can be converted to the active form in the body. In nature, only bacteria can make these compounds. Since they can inhabit the guts of animals, we can get our B_{12} intake from meat. Animals themselves are often supplemented with B_{12} in their feed—in fact, most of the world's production of this vitamin is destined for use as a feed additive to ensure the health of farm animals. Vegans have to rely on supplements.

The cyanocobalamin used in supplements is produced when the product of a bacterial fermentation process is passed through activated charcoal. During this process, it is converted to cyanocobalamin, which is highly stable and suitable for supplements. There is no risk of overdosing.

A study at Tufts University has surprisingly shown that it is not only the elderly who have to be concerned. When three thousand young, healthy adults were tested, 40 per cent had low blood levels of B_{12}, in spite of having a "healthy" diet. It seems that B_{12} from meat, chicken and fish is not as well absorbed as we thought. Perhaps cooking binds the vitamin more strongly to proteins.

Absorption from dairy products is much better. A glass of milk can yield a microgram of B_{12}. Some ·cereals, such as Kellogg's Product 19, are fortified significantly, but others, such as the same company's Corn Flakes, are not. It is a good idea for everyone to make sure they get enough dairy products, and a daily supplement of vitamin B_{12} can't hurt. And take it while you can still remember to do so. B_{12} deficiency will impair memory.

Why are researchers interested in epibatidine, the extremely toxic compound secreted by the Ecuadorian tree frog *Epipedobates tricolor*?

When someone is in pain, every other concern fades into the background. The sole focus becomes alleviating the suffering. That's why pain-relieving medications are so prized. Unfortunately, the most potent ones, the opiates, come with baggage. Their side effects can include confusion, constipation and depression of respiration. Then there is the potential for addiction. So any hint of a discovery of a non-opiate painkiller causes excitement in the pharmaceutical community.

It was back in 1974 that a researcher at the National Institute of Health, John Daly, took an interest in the toxin secreted by the small Ecuadorian tree frog, *Epipedobates tricolor*. South American natives were known to tip their blow darts or arrows with the frog's poison by rubbing them across the creatures' back. "Poison dart frogs," as they have come to be known, synthesize potent toxins from components in their diet, possibly insects, in order to protect themselves from predators. A single lick is enough to teach a predator a very bitter lesson. Daly wondered what sort of compounds the frogs produced and whether they might have some use in medicine. He managed to isolate one specific compound, christened it *epibatidine*, and discovered that in tiny doses it was a very effective painkiller. In fact, in mice it was two hundred times more potent than morphine.

How are such numbers arrived at? It's not as difficult as one might think. Mice, for example, flick their tail in a specific fashion when treated with opiates, known as the *Straub tail response.* The tail-flicking activity is determined by the dose of the opiate administered. So, if the same type of activity is seen in response to a dose of a test drug that is one two-hundredth of a dose of morphine, the drug is said to be two hundred times more active.

Another test used is perhaps more disturbing. Mice are put on a hot plate, and the time it takes for them to jump off is measured. An effective painkiller will lengthen the time. Epibatidine was shown by such tests to be more potent than morphine, and furthermore, the painkilling effect persisted in the presence of naloxone, an opiate antagonist. This suggested that epibatidine was not an opiate and that therefore it might not have an addictive potential.

But Daly's research was hampered by restrictions on the collection of the frogs, which were deemed to be endangered. It wasn't until the 1990s that nuclear magnetic resonance (NMR) technology had developed to the extent that the molecular structure of epibatidine could be determined from the tiny amounts of the compound Daly had collected.

Indeed, epibatidine turned out not to be an opiate; rather, its molecular structure resembled nicotine. It was also one of the rare naturally occurring organic compounds that contained chlorine in its structure. Epibatidine's painkilling activity was related to its ability to activate what are known as *nicotinic acetylcholine receptors* on nerve cells. Unfortunately, epibatidine in slightly higher doses also activates *muscarinic acetylcholine receptors,* causing paralysis and death.

Once epibatidine's structure was determined, chemists went

to work on devising synthetic routes for its production. No longer was there any need to disturb the endangered tree frogs. Alas, the initial hope for epibatidine becoming a non-addictive painkiller quickly faded. The compound was just too toxic. Indeed, it turned out to be more toxic than dioxin, commonly regarded as the most toxic synthetic compound encountered. The therapeutic dose of epibatidine was just too close to the toxic dose for comfort. Although it was a terrific painkiller, it was too likely to kill the patient. But not all hope is lost. As is common practice, pharmaceutical chemists have synthesized a number of molecules closely related to epibatidine with hopes of finding one with a better safety profile. Something may yet dart out of that research.

One of the symptoms of a severe overdose of Aspirin is hyperventilation. Why?

The chemical term for Aspirin is acetylsalicylic acid. As that name suggests, the compound is an acid, and when it is absorbed into the bloodstream from the digestive tract it has an acidifying effect, meaning that it lowers the pH level of the blood. The pH scale measures acidity, with values below 7 indicating an acidic solution and above 7 an alkaline one. Good health requires that the pH level of the blood be maintained in a narrow range, generally between 7.35 and 7.45. To ensure that the value does not wander out of this range, our blood is equipped with a variety of compounds that can neutralize either excess acid or excess base. This makes blood a "buffer" solution, meaning that it resists large changes in pH.

The primary components of the buffer system are carbonic acid and sodium bicarbonate. Carbonic acid in the blood comes about when the carbon dioxide released from cells as they "burn" nutrients to produce energy reacts with water. Any excess base is neutralized by carbonic acid, whereas any excess acid is neutralized by sodium bicarbonate.

The levels of carbonic acid are fine-tuned by the breathing rate. If respiration is very slow, carbon dioxide is not exhaled, carbonic acid builds up in the blood and the pH drops. Hyperventilation, on the other hand, causes loss of carbon dioxide, and since carbon dioxide is needed to form carbonic acid, levels of this acid drop and the pH rises. This is exactly what happens in response to an acetylsalicylic acid overdose. The Aspirin causes blood pH to drop, and in response, hyperventilation kicks in to raise the pH.

An influx of an acid such as Aspirin, or of a large dose of an alkaline substance like baking soda, can affect blood pH, but the effect is temporary due to the blood's rapid buffering action. There is no way to permanently alter blood pH, and in any case this would be highly undesirable because it would lead to severe complications and probably death. The reason for mentioning this is that there are all sorts of claims made by "alternative" practitioners about eating an "alkaline diet" or drinking alkaline water to change the blood's pH in order to prevent cancer. This is based on laboratory experiments that have no relevance to a living body. When cancer cells are maintained in an acidic environment in a test tube, they grow faster, and chemotherapeutic drugs work more effectively in an alkaline environment. These conditions can be set up in a test tube, but not in the body. You cannot alter the acidity of the blood by any sort of diet.

The "alkaline" diet that is talked about is actually an "alkaline ash" diet that can affect the pH of the urine, but not the blood. Whether a food is categorized as acidic or alkaline depends on whether the residue after burning the food is acidic or basic. While an alkaline diet can alter the pH of urine, it does not affect the blood. It isn't a bad diet. It promotes the consumption of fruits, vegetables and nuts at the expense of meat, sugar, alcohol and caffeine, but the claim that this can reduce the risk of cancer is pure folly. There is absolutely no evidence that such a diet can support a sustained change in blood pH, and there is no evidence of any clinical benefit. Testing the saliva or the urine with pH indicator paper is meaningless in terms of offering any clue about blood pH. None of this has stopped a variety of quacks from peddling their water alkalizers and miracle diets to cure cancer. Some even claim to cure diabetes. They make me sick.

The trade name of this drug derives from "diisopropyl intravenous anaesthetic," and it is also colloquially known as "milk of amnesia." What is it?

Most people are unlikely to have heard of the anaesthetic *propofol* until Michael Jackson's unfortunate demise. But the drug became a staple on the nightly news during the manslaughter trial of Michael's physician, Dr. Conrad Murray. The drug goes by the trade name Diprivan, which derives from "diisopropyl intravenous anaesthetic." Propofol is not very soluble in water, but it does dissolve in oil. For intravenous use, it is dissolved in soy oil, which is then mixed with water to form an emulsion. To prevent the oil and water from separating, an emulsifier derived from eggs is added. This is the same idea as adding egg yolk to an oil-and-vinegar mixture to make salad dressing.

A propofol emulsion looks like milk, hence the play on words, "milk of amnesia." Milk of magnesia may settle the stomach, but "milk of amnesia" may lead to settling your estate. Improperly used, propofol can be lethal. Even proper use comes with risk. Of course, the effects of propofol depend on the dose and range from mild sedation to deep sleep. It can be given for short or prolonged sedation as well as for general anaesthesia. When infused, propofol produces a burning sensation that can be countered by the simultaneous administration of the local anaesthetic lidocaine.

Propofol acts very quickly, and recovery usually takes no more than a few minutes after administration of the drug is stopped. Sometimes upon awakening there is temporary memory loss, hence the amnesia reference. While propofol is an excellent anaesthetic, it has to be used with great care because of its potential to depress respiration. Whenever it is used, the patient has to be carefully monitored, and there must be ready access to artificial respiration and cardiovascular resuscitation equipment. Never should this medication be used outside of a medical setting just to induce sleep or to produce a high. Unfortunately, both of these possibilities do occur.

Michael Jackson's death was in all likelihood caused by propofol, probably in combination with other sedatives such as lorazepam (Ativan) that he was taking. Jackson was anxious about his upcoming concert tour, had a very hard time sleeping, and claimed that only propofol worked. It was usually administered to him by Dr. Murray, his personal physician, who maintained that on the night in question Jackson actually self-administered the drug. This was disputed by the evidence offered at the trial, and Dr. Murray was eventually found to be guilty. The fact is that propofol should never have been used as a sleep-inducing agent, and certainly not outside a hospital. Professor Paul Wischmeyer of the University of Colorado, who co-authored a major study on propofol, suggested that using it for sleep was "like using a shotgun to kill a mouse."

Unfortunately, propofol is also abused to attain a "high." A dose can quickly produce a brief euphoric experience and leaves no lingering effects. Use of the drug is hard to detect because it disappears quickly from the bloodstream. Abuse is usually limited to medical personnel, who have access to propofol in a hospital setting and may just slip away for a few moments of bliss, returning to the job minutes later with no sign of having been drugged. But they may not always return. Even with "expert" abuse, deaths are not rare. Estimates are that over a third die from a propofol complication. Obviously, use of propofol doesn't leave much room for error.

BITES
AND SIPS

What alcoholic beverage is referred to as "mother's ruin"?

This term for gin dates back to the eighteenth century, when gin mania in England reached epidemic proportions. Between 1715 and 1750, there were more deaths than births in London, with the greatest mortality among children. Many of these deaths were due to fetal alcohol syndrome, as unhappy mothers-to-be sought solace in gin. And unhappiness was the rule, not the exception, and it wasn't limited to pregnant women. The mid—eighteenth century was a brutal era, rife with robbery, murder and venereal disease. Illness due to a lack of clean water and food was rampant, as smoke spewing from the quickly multiplying factories that ushered in the Industrial Revolution polluted the air.

But gin was cheap and provided at least temporary escape from the abject poverty, filth and hopelessness of the environment. It was the ascension of the Dutchman William of Orange to the British throne in 1688 that marked the beginning of the gin craze. William banned the importation of French wines and spirits and encouraged the distillation of spirits from homegrown grain. Consumption of gin skyrocketed. So did drunkenness and social disorder.

The Gin Act of 1736 attempted to muzzle the runaway gin production by raising taxes on distilled spirits and making the sale of gin in quantities under two gallons illegal. Distillers also had to take out a fifty-pound licence. All this did was cause riots in the streets, lead to prison populations bursting with offenders and stimulate a black market in gin.

As cheap gin flowed unabated, crime increased, men were rendered impotent, women ceased to care for their children, suicide rates jumped, and people sold their possessions to satisfy their insatiable thirst for perpetual drunkenness. All of this was emphatically depicted in William Hogarth's famous 1751 satirical engraving *Gin Lane.* There is the carpenter pawning the tools of his trade for gin, the emaciated dying man still clutching his glass of gin, the neglected infant whose mother is being placed in a coffin, the woman forcing gin into the mouth of an infant to keep it quiet, the schoolgirls drinking gin, a barber who has just hanged himself, and the dominant figure of a woman in a drunken stupor whose child, disfigured by fetal alcohol syndrome, is falling to his death.

Others also took up the battle cry. Henry Fielding ridiculed the alcoholic way of life in *Joseph Andrews* and *Tom Jones.* "What must become of the infant," he asked, "who is conceived in gin with the poisonous distillations of which it is nourished both in the womb and at the breast." Authorities took note that no good was to come of this and passed the second Gin Act in 1751, forcing distillers to sell only to licensed retailers. No longer could gin be purchased from every corner grocer, tobacconist, apothecary, barber or jail keeper. Finally, the gin mania began to fade.

While gin ruined many lives in the eighteenth century, its original purpose was to save lives. It was developed in 1650 as a medicine by Franciscus Sylvius, a professor of medicine at the State University of Leiden in Holland. Juniper berries already had a folkloric history as a remedy for urinary tract problems such as urine retention as well as for gout. Sylvius's idea was to produce a diuretic by distilling juniper berries with spirits derived from fermented barley.

The English term *gin* derives from the French name for juniper: *genièvre*. While the distillate had some success as a herbal medicine, it was its pleasing taste and the appealing effects of alcohol that made it a popular beverage. English soldiers fighting in the Netherlands acquired a taste for the spirit and brought it back home. That precipitated the gin craze, with all its accompanying problems, but also founded the gin distillery industry in England. Numerous varieties were developed, using additives such as oil of turpentine or sulphuric acid for an extra kick, along with lots of sugar to please the palate.

When better stills were developed, allowing some of the bitter fractions of the distillate to be separated out, the need for sugar was reduced and the London dry gin was born. Gin distillers each have their jealously guarded secret recipes, which can include the addition of various botanicals other than juniper berries. Coriander seeds, angelica root, calamus root, cassia bark, lemon and orange peel, licorice and cinnamon are all used.

The final product is incredibly complex chemically, containing hundreds of compounds—in very small doses, of course. Some of these—terpinen-4-ol, for example—have potential biological effects, such as reducing inflammation or stimulating the kidneys' rate of filtration. But there is too little present in gin to have any such effect.

While there is no scientific evidence that gin has any medicinal benefit, one piece of folklore has persisted. That's the use of gin-soaked raisins to treat arthritis. The common recipe is to take a box of golden raisins, soak them in gin for a few weeks until the gin evaporates, and then eat nine a day. Various explanations have been forwarded as to why this works, usually speculating about anti-inflammatory compounds in juniper berries or in the raisins. Pretty far-fetched speculation, given the tiny amounts of these compounds present.

Of course, there are testimonials galore. That isn't surprising since, if you are dealing with a disease that has its natural ups and downs, like arthritis, you can muster a collection of testimonials for anything, be it copper bracelets or snake oil. If an intervention seems to work, you sing its praises, not considering that the improved feeling is just part of a natural cycle. If an intervention does not work, nobody goes around broadcasting the folly of their endeavour.

Famed columnist Paul Harvey's mention of the drunken-raisin phenomenon in 1994 triggered widespread experimentation, resulting in a slew of gushing testimonials. Certainly the most entertaining one came from a correspondent who claimed he couldn't remember if it was to be seven pints of gin and nine raisins or nine pints of gin and seven raisins. He tried both. No pain! My bet is that it was the nine pints of gin that did it, and via the same mechanism with which it can treat a cold. Here's that treatment: when you have a cold, place a hat on the bedpost and start drinking gin. When you see two hats, the cold will be gone. Or at least you'll have forgotten about it.

What are "cured" meats cured with?

Nitrates and nitrites basically define the traditional meat-curing process. In all likelihood, their role was discovered by accident and can be traced to the use of salt that happened to be contaminated with potassium or sodium nitrate, commonly known as saltpetre.

Meat treated with these chemicals retains a red colour, acquires a characteristic taste, and most importantly, is less amenable to contamination with disease-causing bacteria, particularly the very dangerous *Botulinum clostridium*.

By the 1980s it became apparent that certain bacteria were capable of converting nitrates into nitrites and that nitrites were the actual active species. Consequently, nitrites are now added directly to processed meat instead of relying on bacteria to produce them from nitrates. This allows for better control of nitrite concentrations, a critical aspect of processed meat production.

Why critical? Because it is well known that nitrites can react with amines, naturally occurring compounds present in meat, as well as in human tissues, to form nitrosamines. And that is the fly in the hot dog. Nitrosamines can trigger cancer. Of course, demonstrating that nitrosamines can produce mutations in a Petri dish, or that animals treated with high doses develop cancer, does not mean that these compounds are responsible for cancers in humans. In any case, changes in manufacturing methods and a reduction in the amount of added nitrite have essentially solved the problem of nitrosamine formation in cured meat.

In spite of the weakness of the epidemiological evidence linking nitrites to cancer, and the established fact that 95 per cent of all the nitrite we ingest comes from bacterial conversion of nitrates naturally found in vegetables, many consumers have a lingering concern about eating nitrite-cured processed meats. But one person's concern is another's business opportunity. In this case, producers have responded with an array of "natural" and "organic" processed meats sporting catchphrases such as "no synthetic preservatives" or "no nitrites added." But given the crucial role nitrites play in processed meats, how do you replace them? Well, you don't. You just replace the source of the nitrite.

Celery has a very high concentration of natural nitrate, and treating celery juice with a bacterial culture produces nitrite. The concentrated juice can then be used to produce "no nitrite added" processed meat. Curiously, regulations stipulate that the traditional curing process requires the addition of nitrite, and thus "organic" processed meats that are treated with celery juice have to be labelled as "uncured."

Such terminology is confusing because most consumers look to "organic" processed meats in order to avoid nitrites, but the fact is that these *do* contain nitrites, sometimes in lesser, sometimes in greater amounts than found in conventional products. That's because the amount of nitrite that forms from nitrate in celery juice is hard to monitor, while in conventionally cured processed meats the addition of nitrite is strictly controlled by regulations designed to minimize nitrosamine formation and maximize protection against botulism. This means any risk due to nitrosamine formation or bacterial contamination in the "organic" version is more challenging to evaluate.

So, what does all of this mean? Basically, that buying "organic" hot dogs or bacon with a view towards living longer by avoiding nitrites makes no sense. Limiting such foods because of their high fat and salt content, whether organic or conventional, makes very good sense.

What is the primary use of bird nests in China?

For birds to lay eggs in.

Did you say "bird's nest soup"? Well, that certainly is the second most popular use of bird nests. Believe it or not, some two hundred tonnes of nests are consumed in the world every year, with Hong Kong diners leading the way. We're not talking any old bird nest here. These are very special nests constructed by several different species of swiftlets—birds about the size of swallows, found mostly in Thailand, Vietnam, Indonesia, Borneo and the Philippines. The male builds the nest over about thirty-five days, using—I kid you not—his saliva! No twigs, no leaves, just the sticky material secreted from the bird's two sublingual salivary glands.

You wouldn't think that dining on bird dribble would have much of an appeal, but apparently there's no shortage of people willing to shell out up to a hundred dollars for a bowl of bird's nest soup. Why the exorbitant price? Because collecting the nests is a painstaking and often dangerous operation. They're usually found high on cave walls, and harvesting requires climbing specially constructed bamboo trellises while balancing a flashlight between the teeth. Losing one's grip can mean death.

Harvesters also carry a *rada*, a three-pronged instrument they believe was approved by the cave gods. Plucking with bare hands is a no-no, as is making any noise that might disturb the cave spirits.

The traditional method of harvesting has not been able to keep up with the escalating demand for bird nests, prompting some inventive technology. Concrete or wood nesting houses have been built along the sea coast, mostly in Thailand and Indonesia, and seem to be as inviting to birds as the limestone caves. And there are no flimsy scaffoldings or cave spirits to deal with.

After collection comes the tedious process of cleaning the nests. Soaking in water softens the nest cement so that feathers and bits of dirt can be removed with tweezers. The cleaned strands are then moulded into various-shaped pieces of *cubilose*, one of the most expensive food ingredients in the world. How expensive? "The caviar of the East" retails for about $2,000 for a kilo of "white nests" and more than $10,000 for a kilo of "red-blood nests." Cooking a few grams mixed with sugar in a double boiler is said to produce a gastronomical delight. I must admit to being somewhat skeptical about the pleasure that can be derived from sweetened bird spit.

Mostly, though, it's not their palates that people are looking to please, it is the rest of their anatomy. According to traditional Chinese medicine, bird's nest soup imparts a youthful appearance, raises the libido, improves immune function, increases mental focus and treats respiratory ailments as well as digestive problems. And red nests are thought to be more potent than the white.

Given the huge demand for this exotic product, and the difficulty in procuring authentic samples, it comes as no surprise that fraudulent versions have appeared. Leave it to the ingenuity of crooks to come up with adulterations that are difficult to identify. One common technique is to dilute cubilose with less expensive substances such as karaya gum (a tree exudate), a wood-rotting fungus (*Tremella*) or red seaweed. These look, feel and taste like the real stuff, but unlike another adulterant, sodium nitrite, do not raise any safety issues.

Recently, the bird nest industry in China found itself in the midst of a huge controversy after some samples of red cubilose were found to contain more than 350 times the amount of sodium nitrite allowed by health standards. Nitrite is commonly used to add a reddish colour to preserved meats, and according to Chinese authorities may have been used in an attempt to make red cubilose out of the cheaper white version.

There is something puzzling about this, though. Meat contains a complex protein called myoglobin that binds oxygen and releases it as required for metabolic activity. But myoglobin can also react with nitrite to form the red pigment nitro-somyoglobin, the characteristic colour of hot dogs. In the absence of myoglobin, however, nitrite cannot produce a red colour, and the presence of natural myoglobin is unlikely to be found in bird saliva. Possibly, adulteration of nests involves the addition of some sort of meat

extract that has been treated with nitrite. There's something for
the Chinese authorities to examine.

So, what makes authentic red cubilose red? The most likely
answer seems to be iron that leaches out from the walls of caves.
University of Guelph researcher Massimo Marcone has studied
the "caviar of the East" extensively and found that both white and
red nests contain a protein that has properties very similar to a
protein found in eggs called ovotransferrin. This protein forms a
reddish colour when complexed with iron, and Marcone found
that the red nests do indeed contain twice as much iron as the
white. Ovotransferrin is known to be responsible for egg allergies,
and interestingly, very similar reactions have been seen in some
young children after consuming bird's nest soup. It therefore
seems quite likely that red cubilose owes its colour to an iron-
ovotransferrin complex.

Now let's get down to the real nitty-gritty. What, actually, is
cubilose made of, and is it possible that bird's nest soup really has
therapeutic properties? Based on a chemical analysis, it seems
unlikely. Edible bird's nest is 62 per cent protein, 27 per cent
carbohydrate, a few per cent minerals, fat and moisture. Not a
magical combination of nutrients. But nobody has ever run a
proper controlled trial on the benefits of swiftlet spit, and I doubt
anyone ever will. I think I'll stick to chicken soup. Its therapeutic
benefits have not been clearly demonstrated, either, but it tastes
good. And it doesn't cost a small fortune. Still, if someone wants
to treat me to some cubilose, I'm game. Let's stick to the white
variety, though, at least until the Chinese food regulators solve the
adulteration issue.

What did American chemist Wilbur Olin Atwater formulate after burning weighed samples of food?

He came up with the calorie determination system that is widely used, but is actually flawed. Atwater burned weighed samples of food in a *calorimeter* and measured the amount of energy released in the form of heat. After correcting the values for undigested food he isolated from feces, as well as for food metabolites in the urine, Atwater concluded that proteins and carbohydrates provide four calories per gram while fat provides some nine calories per gram. Alcohol weighed in at seven calories per gram, and dietary fibre, which is mostly indigestible, furnished just two calories per gram. These are the values used today to calculate total calorie counts on labels.

But the fate of food in the body is not the same as in a calorimeter. Texture, for example, can make a difference. When rats are fed either soft pellets or hard pellets, they will put on more weight with the soft ones, even if the calorie counts are identical. It is also known that a significant percentage of coarse-ground flour is excreted, while finely milled flour is almost completely digested, yet both would be calculated to have the same number of calories.

Even cooking makes a difference. A well-done steak provides more available calories than a rare one. It may seem that these are insignificant differences, but small differences add up. Making an error in intake of just twenty calories a day can add up to a gain of a kilogram over a year. We need a better way of controlling our food intake than counting calories. And researchers at the Harvard School of Public Health may have put us on track by analyzing surveys filled out by some 120,000 nurses, physicians, veterinarians and dentists over a period of about twenty years.

Starting in the 1980s, these subjects answered questionnaires about their diet and weight, including specifics about the number of servings of various foods they consumed per day. Some fascinating revelations emerged from the massive amount of data collected. First of all, there was an average weight gain of about seventeen pounds over twenty years. Of course, people who exercised put on the fewest pounds. Remember, these were educated people who presumably knew something about nutrition, yet pounds still stuck on, roughly at a rate of a pound a year. The Harvard researchers wondered if there were specific foods that were more likely to cause weight gain.

As one might expect, over the twenty years of the study, there were changes in eating habits, with some subjects eating more of certain foods, less of others. Statistical analysis was able to tease out the effects of these changes on weight.

French fries emerged as the number one villain. People who increased their intake were most likely to put on weight. Other types of potatoes, especially chips, also were linked to weight gain, but not as dramatically as with fries. It may seem that this is obvious, because the fries contain more fat. However, the data also show that an extra serving of nuts a day actually results in the loss of about a fifth of a pound over a year, despite the nuts being rich in fat. So weight control is more complex than just counting calories. Nuts apparently have a greater satiety value and keep people from feeling hungry for a longer time than the fries.

As one would guess, eating more fruits vegetables and whole grains was associated with weight loss. Also as expected, refined grains and, especially, sugar-sweetened beverages were associated with weight gain. It was also no great surprise that red meat and processed meats put on the pounds. But there was a surprise with dairy products. Whether whole fat or low fat, they didn't have much of an effect. The biggest shock was yogurt. People who ate a serving a day were likely to see a weight loss of a third of a pound over a year. That would seem to mesh with some recent research showing that the types of bacteria present in the digestive tract have an effect on the amount of energy that can be extracted from food, and hence on the tendency to put on weight.

What all of this means is that there are foods to stay away from and foods to try to include in the diet. Minimize red meat and eat plenty of fruits, vegetables, nuts and yogurt. Stay away from fries, chips and refined cereals. Sweetened soda pop is the real ogre. The average American consumes half a litre a day. That's about two hundred unnecessary calories! Cutting these calories out will lead to a weight loss, or non-gain, of some two pounds a month.

Here's my take: we need to simplify decision making. So, forget soft drinks and limit fries to once a month—make it a day to look forward to. Those two simple measures are likely to reduce obesity and the associated health costs. Spare me the argument about all foods being all right as long as they are eaten in moderation. In America, "moderation" seems to be a foreign concept.

Why did Americans in the eighteenth century add "pearl ash" to cakes and biscuits?

"Pearl ash" was the common name for potassium carbonate, an alkaline material isolated from wood ash. It served as a source of carbon dioxide gas, which makes dough rise as the gas bubbles try to escape. In order to release carbon dioxide from the potassium carbonate, an acid is required. The classic one used was lactic acid from sour milk. So, recipes called for adding both pearl ash and sour milk to pastry dough. There was, however, a problem with adding potassium carbonate. It reacts not only with sour milk, but also with fats in the food to form soap. And that surely is not the taste we want in cookies.

Eventually, potassium carbonate was replaced by sodium bicarbonate, which we know as baking soda. It has a far smaller tendency to form soap with fats. By 1850, a product had come on the market that contained both baking soda and tartaric acid. When dissolved in batter, the two reacted to release carbon dioxide. Baking powder was born! In order to keep the components dry and prevent them from reacting prematurely, a little starch was incorporated to take up any excess moisture.

But there was still a problem. The baking powder reacted very quickly, and the dough had to be mixed and put into the oven with the utmost speed. Today, baking powder is made with sodium pyrophosphate, an acid that reacts more slowly than tartaric acid at room temperature, but reacts quickly at the high temperatures of the oven.

The comment that "an army marches on its stomach" led to an important discovery in food-preservation techniques. Who made the comment, and what was the discovery?

Napoleon's observation led to the discovery of the canning process. Today's armies are well equipped with food, even on the battlefield, much of it thanks to the availability of all sorts of canned products. But back in Napoleon's time, an army often battled starvation as well as the enemy. Since there were no effective preservation methods, marching armies often subsisted on whatever they could plunder. And hungry soldiers did not fight well, as Napoleon fully realized.

So, in 1800, the emperor offered a prize of 12,000 francs, a great deal of money at the time, to anyone who could invent a practical way of preserving food for his army. Nicholas Appert, who had worked as a brewer, a chef and a confectioner, took up the challenge. Over the years of working with food, he had noted that spoilage occurred more slowly when foods were kept in closed containers. Appert therefore came up with the idea of sealing food in glass jars to prevent contact with the air, hoping for better preservation.

He was disappointed. The contents still spoiled. But he was also aware of the fact that cooked food spoiled less easily. So he had another idea: seal the food in glass jars and then cook it. That worked. Appert sealed his jars with cork, wire and wax before immersing them in boiling water to cook the contents. What sort of contents? Everything from meat and eggs to an entire sheep.

This was no overnight discovery; Appert toiled for over a decade to perfect his process. But by 1811 Napoleon's soldiers were happily feasting on Appert's bottled foods and the emperor, in keeping with his word, awarded the prize to the inventor, who used the money to set up a factory for commercial production. He followed this with a landmark publication entitled *The Art of Preserving All Kinds of Animal and Vegetable Substances for Several Years.*

Though not a particularly catchy title, the paper marked the beginning of the canning industry. It stimulated Peter Durand in England to try to solve the problem of glass jars breaking too easily by experimenting with metal containers. Tubes made from thin sheets of tin sealed at both ends worked well, and the tin can was born. No longer did live animals have to be taken along on long ocean voyages, and Arctic explorations could be undertaken without much concern about food supplies. The can opener not having been invented yet, early cans had to be opened with a hammer and chisel.

Neither Appert nor Durand understood why their method of food preservation worked. The explanation would not be forthcoming until Louis Pasteur demonstrated that spoilage was caused by microbes that could be destroyed by heat, and that preservation could be achieved by preventing further contact of the food with microbes in the air. Ironically, Appert's factory was burned down in 1814 in Napoleon's final battle with Europe's allied forces.

Why are some foods fortified with stanol or sterol esters?

To entice customers who are interested in reducing their blood cholesterol. Elevated cholesterol is a risk factor for heart disease and is not a rare condition. Estimates are that roughly half the adult population has higher-than-desirable cholesterol levels, a condition that shows no symptoms.

Most of our cholesterol is synthesized in the liver, but some originates from eating animal foods. Animals, like us, also make cholesterol. which is in fact an essential biochemical. It is a part of all cell membranes and is the raw material used to make important compounds such as the sex hormones. Plants do not produce any cholesterol, but they do make closely related compounds called sterols and stanols. Because of their chemical similarity to cholesterol, it turns out that sterols and stanols can compete with cholesterol for attachment to transport molecules that ferry cholesterol from the gut into the bloodstream. As a result, less cholesterol is absorbed from the digestive tract, and blood levels drop.

Sterols and stanols are found in a wide variety of foods, with whole grains leading the way. But the total amount consumed in the average diet is only about 300 milligrams, and that is not enough to significantly reduce blood cholesterol. Numerous studies have shown that to have an appreciable impact, a daily dose of roughly 2 to 2.5 grams is needed. That can lower blood cholesterol by 10 to 14 per cent, an amount that can significantly reduce the risk of heart disease. Since obtaining two grams of sterols and stanols from the diet is not possible, a whole industry based on fortifying various foods with these compounds has arisen.

Industrially, they can be isolated from soy, corn or various types of pine trees, after which they are combined with fats, mostly canola oil, to afford a consistency suitable for adding to foods such as margarine or yogurt. Four tablespoons of fortified margarine a day can deliver an adequate amount of sterols and stanols, but that's a lot of margarine. Yogurt seems to be a better vehicle for delivery. Indeed, just two 100-gram servings a day can meet the guidelines for cholesterol lowering.

Of course, another question needs to be asked. Is there any risk associated with pumping that much stanol or sterol esters into our bodies? After all, it is not an amount that can be consumed naturally. According to the long-term clinical studies that have been conducted, there is no risk to such an intake. But neither is there evidence that people who include foods fortified with stanols or sterols have a reduced risk of heart disease.

✯

What vegetable is grown in a way that prevents photosynthesis?

White asparagus is grown covered in soil to prevent the stalks from turning green as a result of photosynthesis. This results in a sweeter and more tender product, particularly prized in Europe. *Spargel,* as the Germans call white asparagus, is harvested each spring, mostly around the town of Schwetzingen, the "Asparagus Capital of the World." The area even has an annual Spargelfest that features the naming of a Spargel Queen, undoubtedly a much sought-after honour.

Asparagus is a fascinating vegetable in many ways. It can taint the urine with a characteristic aroma, but there is controversy over whether or not everyone, or just some genetically predisposed people, produce the fragrance. To make things even more interesting, the ability to smell the asparagus taint may also be a genetic trait. There is no controversy, though, about the speed with which asparagus grows. Fast! A spear can grow as much as ten inches a day. Given this amazing growth rate, and the resemblance of a spear to a part of the male anatomy, it isn't surprising that in some cultures asparagus is regarded as an aphrodisiac. There is absolutely no evidence for such an effect, nor for the widely circulating Internet claim that asparagus can cure cancer.

The "asparagus for cancer" nonsense comes from an unidentified "biochemist" who supposedly teamed up with Richard Vensal, a dentist who does not appear to exist outside of the circulating email, to publish an article in *Cancer News Journal,* a rag that cannot be found in any library. The mysterious duo describe a string of miraculous cures linked to the daily consumption of eight tablespoons of pureed asparagus. The "biochemist" in question goes on to suggest that "what cures can prevent" and informs us that he and his wife concoct a beverage to complement their breakfast and dinner by diluting two tablespoons of the puree in a glass of water. He attributes the wonders of asparagus to proteins called *histones.* More flagrant nonsense.

All animal, plant and fungal cells contain histones, which are the proteins that act as a sort of spool around which DNA in each cell's nucleus is wound. They do play a role in the activation of genes, and therefore do have an effect on cell growth, but that in no way means they have an anti-cancer effect. Of course, the point is moot anyway because histones, like other dietary proteins, are broken down during digestion, so no amount of asparagus, or any other vegetable, fruit or meat, can pump histones into the nucleus of a cell. There is no such thing as a histone-deficient diet.

A final point made by the wise biochemist is that asparagus is an outstanding source of glutathione, which is one of our body's most important detoxicating compounds. He is correct in saying that asparagus is an excellent source of glutathione; in fact, it contains more than any other vegetable. He is also right about glutathione performing a number of important detoxicating functions in the body. But he needs to review his biochemistry. Glutathione is made of three amino acids linked together and is immediately broken down into its components when it hits the acid environment in the stomach. You cannot increase your glutathione levels by eating foods rich in glutathione.

None of this is to suggest that asparagus cannot play a role in cancer prevention, just like any other vegetable. There's plenty of evidence that consuming vegetables can have a protective effect, although the exact mechanism of the protection is not clear. That's because vegetables contain a huge variety of compounds that, in laboratory experiments, can be shown to have all sorts of potential anti-cancer effects.

Ferulic acid, found in asparagus, can inhibit the growth of tumour-supporting blood vessels; quercetin and rutin can destroy some cancer cells; and some steroidal saponins have been shown to promote the selective death of liver cancer cells. Intriguing laboratory findings, but we can only guess at what they mean in terms of humans eating asparagus. Probably not much. Asparagus also contains a number of carotenoids, like beta carotene, lutein and zeaxanthin, which have been linked with a reduced risk of cancer in population studies, and the purple-stalk variety features anthocyanins, the same pigments found in blueberries. Anthocyanins also have a reputation as anti-cancer substances because they are potent free-radical scavengers, at least in the test tube.

We've all heard the saying that "what happens in Vegas stays in Vegas." Well, what happens in the lab may just stay in the lab. Demonstrating some biological activity in a Petri dish does not mean that the same activity occurs in the body. Suggesting that we eat asparagus as part of a healthy diet is fine, but any suggestion that asparagus can cure cancer stinks. And I can smell that as well as I can smell asparagus taint in the urine.

What is the supposed benefit of taking a dietary supplement made from the seeds of the African mango?

Obesity is one of North America's greatest nutritional challenges, and it is now creeping into the developing world as well, thanks to the export of our dietary habits. But maybe we can import a weapon from another culture to help us in the battle against the bulge.

That weapon is an extract of the seeds of *Irvingia gabonensis*, a West African fruit that goes by the common name of African mango. Whether this extract will end up being embraced by the scientific community as an effective way to control weight or end up on the junk heap of weight-loss drugs and dietary supplements that failed to meet the initial hype remains to be seen. At this point, all we can say is that, based on an intriguing study carried out in Cameroon, further investigation is warranted.

Just what generated the idea that the African mango should be linked to weight control isn't clear, but it was probably a combination of looking for new markets and a finding that the seeds were high in soluble fibre. There is some evidence that soluble fibre delays stomach emptying, which in turn reduces appetite. Furthermore, an intake of soluble fibre is known to reduce blood cholesterol and is also known to reduce the elevation of blood sugar after a meal, which is handy when it comes to controlling diabetes. A small double-blind randomized study of forty subjects in 2005 showed a significant weight loss of about four kilograms with three grams of the extract taken daily. There was also a decrease in cholesterol and triglycerides, as well as in waist and hip measurements. But there are some peculiarities in the data.

First of all, the average initial weight of subjects in the experimental group was 105 kilograms, while that in the placebo group was 79 kilograms. In a sense, we are comparing apples to oranges here, because any sort of intervention will cause a greater initial weight loss when there is more weight to lose. Why were the subjects not randomized in such a way that the initial weights in the experimental and control groups were roughly equal?

A later study in 2009 eliminated this concern, with the initial weights of the two groups being essentially equal. This time, there were 102 volunteers randomly divided into two groups, with the experimental group swallowing a capsule containing 150 milligrams of African mango extract an hour or so before lunch and dinner. Why only one-tenth as much of the extract as in the 2005 study was used isn't addressed in the paper published in *Lipids in Health and Disease,* a peer-reviewed but not top-of-the-line journal.

It is likely that the extract, provided by a company that markets the product, was a different version than that used in the original study. The weight loss after ten weeks, with one-tenth as much of the extract, was about 13 kilograms, far more than in the original trial, as was the 16-centimetre decrease in waist size. Curious. There was also an astounding drop in total cholesterol of 40 per cent and blood glucose of 19 per cent. These are amazing and surprising numbers in comparison with what has been seen in other weight-loss regimens.

How do the researchers explain these stunning results? By invoking changes in certain hormone levels that they have measured. Leptin goes down and adinopectin goes up—both effects have been linked with weight loss. But before jumping on the bandwagon, the one where TV's Dr. Oz has already planted himself in the driver's seat, I'd like to see some replication of these results. Just sounds too good to be true.

Where would you find "brominated vegetable oil"?

There's no point searching for it in the cooking oil section of the supermarket. But if you stroll down the beverage aisle and pick up a citrus-flavoured drink, you are likely to find *brominated vegetable oil* listed on the label as an ingredient. Why is it there? To distribute citrusy flavours that are normally not soluble in water throughout the beverage and, as an added bonus, to provide a cloudy appearance reminiscent of real fruit juice. The citrus flavour that is destined for use in drinks is a complex mix of compounds extracted from the rinds of citrus fruits. Some of these are not water-soluble, meaning that they would either rise to the top or sink to the bottom if the extract were used to flavour a beverage without any further chemical manipulation. It is during this further manipulation that brominated vegetable oil comes into the picture.

The citrus flavours extracted from the rind that do not dissolve in water do dissolve in oil. But, of course, that would not be of much help in concocting a beverage because the flavoured oil, being immiscible with, and lighter than water, would just separate and rise to the top. However, adjusting the specific gravity of the oil to match that of water allows the oily flavour droplets to be evenly distributed through the drink. True, the oil still remains immiscible with water, but that is not regarded as a detriment. Quite the opposite: the resulting cloudy appearance conjures up images of fruit juice.

But how do you increase the specific gravity—in other words, the weight of the oil—to achieve this desired effect? You brominate it. "Unsaturated" vegetable oils, such as soy or corn oil, contain carbon-carbon double bonds that readily react with bromine, a brownish liquid in its elemental state. Bromine atoms are heavy, so once they become incorporated into the structure of the vegetable oils, they increase the molecular weight to a degree that the oil has roughly the same specific gravity as water.

Needless to say, there is controversy about the use of brominated vegetable oils. Add anything to a food or beverage that is seen to be "unnatural," and you are guaranteed to hear some clucking about how that chemical is poisoning us. Indeed, an excessive intake of bromine in virtually any form can be toxic. The emphasis, though, is on "excessive."

Amazingly, until 1975, potassium bromide in doses of several grams a day was used as a sedative. Seems strange that, till then, nobody had connected taking bromides with an unusually high rate of admission to psychiatric facilities. Indeed, *bromism* can cause confusion, hallucinations and even psychosis. Add to this neurological abnormalities, memory impairment and gross skin pustules, and you've got a nasty condition. But it takes a lot of bromine to cause such misery. Certainly, taking grams of potassium bromide can do it, but beverages? Not unless you drink about eight litres a day. Sound crazy? Well, it has happened.

A case report in the *New England Journal of Medicine* describes a sixty-three-year-old man who presented with terrible red ulcerated nodules on his hands. He volunteered the information that he had been drinking eight litres of Ruby Red Squirt daily for several months. Eliminating the beverage reversed the *bromoderma*. Surprisingly, the medical literature records another case of a man who saw his physician because of increasing headaches, fatigue, balance problems and memory loss. The doctor was stymied as his patient deteriorated until he was unable to walk. Eventually, a blood test revealed a stunningly high level of bromine in his blood, at which point the patient admitted to drinking two to four litres of a soft drink formulated with brominated vegetable oil daily. Luckily for him, dialysis was able to clear his blood of bromine, and he managed to recover.

The moral of the story is that consuming litres and litres of any liquid every day is not a good idea—not even if it is water. Undoubtedly, though, the scare stories about brominated vegetable oil will continue on the web unabated. One brilliant mind offers the opinion that brominated vegetable oil is processed sludge from the bottom of the ocean. Where do people get such ideas? Well, bromine is, in fact, isolated from seawater. Of course, that has nothing to do with evaluating the health effects of brominated vegetable oil. Maybe that brilliant mind is filled with sludge.

☆

Where would you go to look for guarana?

If you wanted to find guarana, you would trudge through the forests of the Amazon basin, keeping your gaze on the trees, looking for eyes that look back at you. The Amazon is a dangerous place, with jaguars, anacondas and piranhas in constant search for their next meal, but you would not have anything to fear from guarana. It isn't a predator; it's a woody vine that climbs through the trees, growing up to thirty feet long. It produces bright red berries that split open when ripe, revealing a shiny black seed partially embedded in a thin white pulp. From a distance, the split berries look disturbingly like eyes staring down from the leafy canopy. The name *guarana* reflects this connection, deriving from the native words *guara*, for "human," and *na* for "like."

Amazonian natives had discovered the invigorating effects of guarana long before Europeans ever arrived in South America. Traditionally, they ground the dried seeds into a paste that was then formed into sticks. Grating the dried sticks into water produced a beverage that was mainly used as a stimulant. And did it ever stimulate! No great surprise there, for the simple reason that guarana is a rich source of caffeine and also contains theobromine and theophylline, stimulants very similar to caffeine. Sometimes, the three stimulants are collectively referred to as guaranine.

Weight for weight, guarana seeds contain about twice as much caffeine as coffee beans. Indeed, in South America, most of the caffeine used in beverages and pills is extracted not from coffee beans, but from guarana seeds. Extracts of the berries are also used to flavour soft drinks that are extremely popular in Brazil. Besides looking for a stimulant effect, Brazilians also consume guarana in the form of candies, syrups and various herbal tonics with hopes of "purifying the blood," preventing premature aging and reducing the appetite. While caffeine does have an appetite-suppressing effect, the other claims are pure whimsy.

Guarana has now wended its way to North America, where it is most likely to be found as a component of "energy drinks" that claim to provide consumers with, what else, extra energy. That energy comes from a good jolt of caffeine, a hefty dose of which comes from guarana, but carnitine, glucuronolactone, inositol, ginseng, hydroxycitric acid, taurine and yohimbine are also commonly added in various combinations, with claims ranging from improving endurance and suppressing appetite to boosting sexual performance and promoting the excretion of toxins.

The stimulant effect of caffeine is very real, but the other claims lack scientific evidence. While modest amounts of caffeine do not pose a problem, and may indeed help with mental alertness and exercise endurance, some of the energy drinks contain as much as 300 milligrams of caffeine per bottle. That's more than double the amount found in the strongest cup of coffee. Such a dose can not only make one jittery, it can affect blood pressure and cause palpitations and, in rare cases, even seizures.

Doctors at St. Joseph's Hospital and Medical Center in Phoenix report a series of four young patients who had seizures on multiple occasions following heavy consumption of energy drinks. Once they were told to abstain from the beverages, the seizures ceased. Because the drinks contain multiple components, it is not possible to say with certainty that caffeine is the culprit, but it is the most likely candidate. The seizures were most common when the drinks were consumed on an empty stomach.

There is also concern about consuming guarana in any form together with alcohol. Guarana, because of its high caffeine content, produces a sensation of alertness that may cause imbibers to overlook the debilitating effects of intoxication. A recent study reports that consuming alcohol with energy drinks leads to a reduced feeling of intoxication while having no effect on objective measures of motor coordination and reaction time. This means that someone who has consumed alcohol and guarana may feel that they can drive because they don't feel drunk. But they may well be. Guarana berries may look like they're looking at you, but they are not looking out for you.

Why should you not eat polar bear liver?

The simple answer is that it could kill you. Death would come from an overdose of retinol. That isn't some bizarre toxin; retinol is just the chemical name for vitamin A.

Polar bears are carnivores, feeding mostly on seals. Lots of seals. Their hearty appetite is due to their need to accumulate lots of fat for insulation against the cold. Seals are rich in blubber. And where does their fat come from? They feed on fish, squid, krill and other sea creatures, which in turn feed on various types of plants ranging from seaweed to algae. These plants contain various carotenoids that concentrate up the food chain and serve as precursors for vitamin A. Since vitamin A is fat-soluble, it builds up in fatty tissue, particularly the liver. Polar bears have lots of fatty tissue, so they accumulate lots of vitamin A, particularly in their livers.

Like virtually any substance, vitamin A can be toxic in high doses. But polar bears have evolved to be unaffected by the higher doses of vitamin A, because its accumulation is a consequence of their need for a high-fat diet. Humans do not require the same degree of insulation, so we have not had the need to eat as much fat as polar bears, although some people do give it a gallant effort. A blood level of vitamin A that does not bother a polar bear can kill a human. And it has. Arctic explorers have died when, due to a scarcity of food, they had to resort to hunting polar bears.

Vitamin A is a classic example of a substance that is beneficial in small doses and toxic in higher doses. It is required by the body for a variety of purposes. It plays an important role in vision, bone growth, reproduction, immune activity, cell division and cell differentiation, the process by which a stem cell becomes a brain, muscle, lung, blood or other specific cell. But amounts are important. While vitamin A is needed in small doses for all these bodily processes, in larger amounts it interferes with them. The result can be blurred vision, loss of hair, infections, birth defects and death. Outside of consuming polar bear liver or going crazy with supplements, overdose of vitamin A is not an issue. Underdosing is a greater concern.

There are two possibilities for meeting our vitamin A needs. Meat contains vitamin A, with liver containing the most. So do eggs and milk. Most nonfat milk products are fortified with vitamin A to replace the amount lost when fat is removed. Some cereals are also fortified. But it is not necessary to eat animal or fortified products to meet our vitamin A needs. We can do so by eating plant products, in spite of the fact that no plant contains vitamin A. That seems to be a conundrum. While it is true that plants have no vitamin A, they do contain compounds called carotenoids that the body can convert to vitamin A. Beta-carotene, found in the likes of carrots, spinach, cantaloupe, papaya, mango and oatmeal, is a common precursor for vitamin A.

And you do not have to worry about overdosing on vitamin A by eating too many carrots. When enough vitamin A has been stored in our livers, our bodies stop converting beta-carotene to retinol. The worst that can happen with a carrot overdose is that the skin can turn a yellow-orange colour. Fortunately, it's reversible.

What is xylitol doing in your chewing gum?

Giving you a sweet experience without harming your teeth. Hopefully, it does this without sending you scurrying to the bathroom.

Sugar, as we well know, is persona non grata as far as our teeth are concerned. Bacteria in our mouths find sugar delicious, and they happily ingest it. But like us, bacteria also poop. And when they consume sugar, they poop out acids that can corrode the tooth's enamel and cause cavities to form. But chewing gum that isn't sweet isn't much fun. Artificial sweeteners like aspartame can be used to impart flavour, but xylitol can do the job as well. The bacteria responsible for cavities cannot metabolize xylitol and therefore cannot multiply at the same rate as when fed sugar. In high doses, more than about 15 grams a day, xylitol can cause stomach problems, and at even higher doses, it can give you the runs. Obviously, the dose matters both in terms of benefit and risk.

Since streptococcus bacteria can be transferred to a baby through a mother's kiss, xylitol-sweetened gum is especially appropriate for new moms. Studies show that transmission of bacteria during the first two years of life can be reduced by as much as 80 per cent! Xylitol's safety has been thoroughly tested, including in pregnant and nursing women, and aside from a laxative effect at doses way above what people actually consume, no effect has been noted. Xylitol has approximately the same sweetness as sugar, but only contributes about 2.4 calories per gram as opposed to sugar's 4 calories per gram. Furthermore, xylitol does not need insulin to be metabolized, so it can be safely consumed by diabetics.

A common question that comes up about xylitol is whether it is or isn't an artificial sweetener. This usually arises because of concerns about "artificial sweeteners." Essentially, the question is an inappropriate one, because a compound's safety profile is not determined by its origin. Whether xylitol is extracted from some berry, which it can be, or whether it is made by an industrial process from xylose, which it usually is, has no bearing on its properties. The reason it has been found to be safe is simply because it has been tested. Xylitol is xylitol no matter where it comes from.

So, is it an artificial sweetener or not? I guess it is really both. It does occur in nature—you'll find it in corn husks, fruits, berries, oats and various trees. But it isn't practical to isolate it from these sources. On the other hand, xylose, a common carbohydrate, can easily be produced from xylan, a type of fibre found in the cell walls of many plants. The pulp and paper industry is a ready source for xylan, which in turn can be converted to xylitol through a hydrogenation process. And then it's ready to be added to gums, candies, toothpaste, pharmaceuticals and mouthwashes.

But keep these away from dogs. Dogs metabolize xylitol differently, and at high doses can exhibit seriously low blood sugar and even liver damage.

What is liquid smoke used for?

Subjecting various foods to smoke is an age-old method used both for preservation and flavouring. But there are issues. Smoke contains a variety of cancer-causing compounds. And in this case, we are not talking about a hypothetical cancer risk to humans based upon feeding some outrageous amounts of a suspected carcinogen to rodents. We have robust evidence that smoke can cause cancer in humans.

It was in 1775 that British surgeon Sir Percival Potts discovered that chimney sweeps had an unusually high incidence of skin cancer on their scrotums. It was also noted at the time that chimney sweeps on the continent did not experience this rare cancer. What was the difference? Bathing was more common on the continent than in Britain, and British chimney sweeps walked around continually covered in soot from head to toe. Maybe there was something in soot that was responsible for the disease. The gucky material inside chimneys, known as creosote, was the likely candidate. Scientists began to explore this possibility by painting creosote on the skins of animals to see if tumours could be induced. For about 150 years, these experiments went on sporadically and unsuccessfully.

Then along came Katsusaburo Yamagiwa, a medical researcher in Japan. He had something that his predecessors in this research area apparently did not have: patience. Yamagiwa carefully painted the ears of 137 rabbits every day for over a year. Finally, tumours began to appear on seven of the rabbits. This could not happen by chance; Yamagiwa had found a carcinogen. Actually, he found many carcinogens. Creosote is a very complex mixture of chemicals, and to this day it has not been completely analyzed. But some of its components, like the notorious polycyclic aromatic hydrocarbons, or PAHs, have been identified as definite carcinogens. They, however, are not found in liquid smoke. Now we can have our smoke flavouring practically free of any worrisome carcinogens.

Since the 1970s, millions of gallons of liquid smoke have been sold to impart flavour and aroma to foods. It is the natural aqueous condensate of wood smoke that is aged and filtered to remove tars and particulate matter. Smoke from the combustion of wood rises through a condensing tower, where it mixes with cold water. The mixture is allowed to settle for ten days in a storage tank, where insoluble compounds such as benzopyrene settle out. This is followed by a multistage filtration process.

More than four hundred compounds have been identified in liquid smoke, their presence depending on the type of wood burned, its age and growing conditions. Components include acetic, propionic, butyric and valeric acids that affect the flavour, pH and shelf life of the product; carbonyl compounds that react with proteins to produce colour; phenols that contribute flavour, bacteriostatic activity and antioxidant properties. And there are virtually no polycyclic aromatic hydrocarbons. In this case, the processed product is better than the natural.

What agent is used to make cola beverages their brown colour?

Colas derive their colour from caramel, which basically is a complex mixture of compounds produced when various carbohydrates such as sucrose, fructose, glucose or starches are heated to a high temperature. Put some sugar or starch in a pan, heat it, and soon you'll have caramel. And, of course, a mess to clean up. Chemically, caramelization is a very complex process resulting in the formation of hundreds of compounds. To complicate things further, there are several types of commercial caramelization processes, depending on what other reagents are added to the carbohydrate source as it is being pyrolyzed. Adding acids or alkalis to promote caramelization is common, but it is also possible to add sulphites, such as sodium sulphite, or ammonia compounds, such as ammonium carbonate, to achieve specific shades of brown.

Caramelization also occurs commonly in cooking and produces basically the same set of compounds as a commercial process, except for when ammonium compounds are added. Any complex mixture like this will contain some compounds that, when isolated and carefully investigated, will produce some adverse effects in cell cultures or laboratory animals. When ammonium compounds are used, some of the breakdown products fall into the imidazole family, a couple of which—namely 4-methyl-imidazole and 2-methylimidazole—have been shown to cause cancer in rats and mice. This can provide some grist for the alarmist mill. And it has.

The Center for Science in the Public Interest, usually a reliable organization when it comes to nutritional issues, has overstepped the boundaries of good scientific sense in its petition to the U.S. Food and Drug Administration to have caramel produced by the ammonium addition process banned as a food additive. It claims that the imidazoles in cola colouring could be responsible for thousands of cancers a year. What evidence is there for this? Aside from two studies using isolated imidazoles that showed carcinogenicity in rodents, none. There are all sorts of studies showing that caramel colouring itself is not carcinogenic. Numerous examples exist of mixtures that are not carcinogenic despite the presence of some carcinogens. Coffee is a classic example. No regulatory agency has listed caramel, or indeed any of its components, as a carcinogen. Save one: California!

The state has proposed that 4-methylimidazole be added to the list of "chemicals known to the state of California to cause cancer." That would mean all cola drinks in California would have to carry a cancer warning, according to the state's bizarre Proposition 65, which states that any product that contains a carcinogen requires a warning label. Caramel colouring is also used in beer, bread, chips, doughnuts, ice cream, whiskey and a myriad of other items, so supposedly all these would require warnings as well. This is not sound science. To equal the amount of imidazole the rats were fed over a seventy-two-hour period, a human would have to consume roughly 12,000 bottles of cola.

European regulatory agencies do not buy into this fear-mongering. The European Food Safety Authority (EFSA), certainly not known for its laxity when it comes to evaluating safety, has carried out a reevaluation of caramel colouring and has concluded that caramel colours are not carcinogenic. EFSA has established a safe intake of caramel colouring at 300 milligrams per kilogram of body weight. Nevertheless, many cola producers, wanting to avoid legal hassles in California, are altering their caramel production methods to eliminate methylimidazoles.

A butcher buys a bag of white powder labelled *Transglutaminase*. What does he propose to do with it?

The idea of eating a steak made from pieces of meat scrap glued together with transglutaminase is likely to stick in most people's craws. But there are also those who are ready to shell out a small fortune at New York's WD-50 restaurant for a chance to sink their teeth into "shrimp noodles" concocted with the same "meat glue."

So, what is this 'meat glue'? Rest assured that no horses were condemned to the glue factory to produce it. What we're talking about is an enzyme called transglutaminase that allows a mouthful of shrimp to be served in the form of noodles that look, but certainly do not taste, like regular pasta. How does it do this? By facilitating a chemical reaction that forges links between structural protein molecules. Proteins, of course, are composed of chains of amino acids, and transglutaminase links the amino acid lysine in one chain to glutamine in an adjacent chain. If these chains are located on the surface of adjacent pieces of meat, the pieces get stuck together almost like magic. The joint then looks just like one of the white streaks of gristle or fat commonly seen in meat. It's so strong that the meat doesn't even tear along the "fault line."

Transglutaminase is not foreign to the human body. We produce it to aid in blood clotting, a process that requires protein molecules to form interlinked complex structures. Skin and hair are also composed of proteins that have been bound together, and transglutaminase plays a role here as well.

In the 1990s, the food industry discovered that this enzyme could be isolated in good yield from the bacterium *Streptoverticillium mobaraense* and that it can be used to "restructure" meat, fish and poultry. For example, with the help of transglutaminase, bits of chicken left over after the carcass has been processed, instead of being discarded as waste, can be glued together to produce chicken patties. Similarly, artificial crab legs and shrimp can be made by sticking together ground pieces of cheaper seafoods such as pollock. While the taste of such artificial foods can be criticized, there is no health issue associated with eating transglutaminase. Like any other protein, it is readily broken down into its component amino acids in the digestive tract.

"Meat glue" is produced for the food industry under the name Activa by the giant Japanese company Ajinomoto, which also markets monosodium glutamate (MSG). It did its work quietly behind the scenes until celebrity chef Heston Blumenthal brought it out of the shadows at the Fat Duck, the restaurant on the outskirts of London that has been labelled by many as the best eating place in the world. Blumenthal's enthusiasm for creating novel dishes with transglutaminase rubbed off on Wylie Dufresne at New York's famed WD-50 restaurant, who managed to grind shrimp into noodles with the help of transglutaminase and served it on a bed of smoked yogurt.

Other chefs who pursue what has been called *molecular gastronomy*, defined as the application of scientific principles to the creation of new dishes, are pushing the transglutaminase envelope. Around the corner are filet mignon with strips of bacon glued to its surface, fish coated with chicken skin to enhance flavour, and shrimp burgers held together by cross-linked proteins. How about chicken fat stuck to steak to add a new dimension to chicken-fried steak? Just in case your cholesterol isn't high enough.

Unfortunately, transglutaminase also lends itself to some less savoury applications. Such as producers or butchers using it to bind meat scraps too small to be sold into slices that look every bit as delectable as prime cuts. The seamless joints are virtually undetectable. Ditto for any difference in taste. Just take the bits of meat, sprinkle them with transglutaminase, place them on a sheet of plastic wrap and roll tightly into the shape of a tube. Refrigerate for a few hours and then unwrap. You'll be looking at meat that, for all the world, looks like a single piece of filet. And it can be priced accordingly.

Clearly, there is deception involved here. The customer is not getting what he or she is paying for. There are some other issues as well. Transglutaminase can be isolated from blood, with bovine and pig blood being used commercially. This can be a problem for people adhering to religious dietary laws. Not only can transglutaminase be used to make "restructured meat," it can also be used to improve the texture of hot dogs and sausages. Meat glue is not allowed in Europe, but can be used in Canada as long as it is declared on the label like any other additive. For example, if Chicken McNuggets were glued together with this enzyme, it would have to be listed on the ingredients list, which is available from McDonald's. It's not. So, no "meat glue" there. Whether or not some unscrupulous butchers use it to make fake steaks is another matter.

Any butcher engaging in such clandestine operations may pay a price. While ingesting transglutaminase is no problem, inhaling the powder can damage the lungs. Consumers don't have to worry about this, but there is an issue with cooking glued meat. The surface of meat is always covered with bacteria, but the microbes are readily killed by cooking. However, with structured meat, some of the outside becomes the inside, and if the meat is not thoroughly cooked—as is clearly possible for people who like their steak rare—bacteria on the inside may survive. This is the reason why hamburger meat, a classic "outside becomes inside" situation, has to be cooked through and through.

What if meat glue gets on the hands? No sticky fingers here—contact time is not long enough to do anything. The only sticky fingers are the ones involved in extracting money from people by passing off glued scraps as prime cuts. Finally, I wonder if Lady Gaga's famous meat dress was held together with transglutaminase. Only her butcher knows for sure.

To what does Peking duck owe its traditional red colour?

Peking duck is a classic Chinese dish characterized by crispy reddish skin. Traditionally, the colour was achieved by rubbing the duck with red yeast rice, a fermented rice product known in China for over a thousand years. The mould *Monascus purpureus* that is used to inoculate the rice produces a number of compounds in the *polyketide* family with hues ranging from orange to red, suitable for colouring a number of food products. Besides Peking duck, products including Chinese sausages, soybean products, rice wine and rice vinegar are often dyed with red yeast rice.

Perhaps because of its red colour, this fermented rice has also been used in traditional Chinese medicine as a "blood revitalizer." That may actually have some meaning if one interprets "blood revitalization" as a lowering of blood cholesterol. In the 1970s, a compound christened monaculin K was isolated from red yeast rice and found to inhibit cholesterol synthesis in the liver by interfering with the activity of HMG-CoA reductase, an enzyme critical to cholesterol synthesis. Before long, monaculin K was found to be identical with lovastatin, the first-ever cholesterol-lowering statin, which had been isolated from an aspergillus fungus by researchers at Merck.

As Mevacor, lovastatin became a patented prescription drug generating huge sales. Red yeast rice went on to become a "natural" alternative to prescription statins, enthralling those who believe that natural products are inherently safer than anything synthetic. Of course lovastatin is lovastatin, no matter where it comes from. Whether the compound is produced synthetically in the lab or isolated from a mould is irrelevant.

Red yeast rice actually does work to lower LDL, the so-called "bad cholesterol," which is not surprising since it contains a statin. And lowering LDL is desirable, given that studies have shown a major impact on major coronary events. Red yeast rice's ability to reduce cholesterol has been shown in a number of clinical studies, and patients treated with the supplement were found to have not only reduced cholesterol levels but also lower triglycerides. Furthermore, there was a decreased risk of heart attacks as well as deaths from heart-related causes. The red yeast rice was well tolerated and side effects, such as muscle pain, were no greater than with placebo. The amount of lovastatin in the red yeast rice that was used was only about 10 milligrams per day, which is half of the smallest usual prescribed dose of pure lovastatin, so the low side-effect profile is not surprising.

So if red yeast rice actually works, why is it that the Food and Drug Administration in the U.S. prohibits the sale of red yeast rice that contains any but a trace amount of monaculin, and Health Canada has not licensed any red yeast product? It isn't because red yeast rice can't lower cholesterol; it is because, due to soft regulations on natural health products, the consumer cannot count on any specific preparation to contain a standardized amount of lovastatin and cannot count on the product to be devoid of citrinin, a mould byproduct that can cause kidney damage.

Even if a red yeast rice product that contains no contaminants and a meaningful amount of lovastatin can be found, it still presents some problems. Lovastatin, whatever its source, can cause side effects, most notably muscle aches, and in rare cases rhabomyolysis, a serious breakdown of muscle fibres. Liver function can also be affected, which is why patients taking statins have to undergo periodic liver enzyme tests. If someone is self-treating with red yeast rice, such problems may not be picked up until it is too late.

The pertinent question is not whether red yeast rice lowers cholesterol, but whether use of this "natural" product is safer than a prescribed statin. It isn't. The active ingredient is a statin, but there is no guarantee of amount or purity. As with many such "natural" products, you don't know what you are getting. In any case, since neither the FDA nor Health Canada allows sales of red yeast rice with any significant amount of statins, any product the consumer is likely to buy will be ineffective in reducing cholesterol.

As far as Peking duck goes, there is no need to be concerned about the amount of lovastatin. It would be trivial. Interestingly, this dish played a role in the rapprochement between the Chinese and American governments in the 1970s. Henry Kissinger was introduced to Peking duck on his first visit to China and loved it. He jokingly suggested to premier Zhou Enlai that it was too bad U.S. President Nixon could not enjoy this dish. The following day, the premier and Kissinger issued a joined statement inviting Nixon to visit China, which he famously did in 1972.

What are researchers trying to do when they promote the multiplication of pig stem cells in a nutrient solution of horse fetal serum?

Most people agree that meat tastes good. But the idea of raising animals and killing them for our gustatory delight leaves a bad taste in the mouth. Wouldn't it be great if we could produce steaks, burgers and sausages without killing animals? Sound like science fiction? Well, it may be more science than fiction. Researchers are on the verge of producing "in vitro" meat, meaning meat that is produced without the need for an animal. Actually, that is not completely true, because the stem cells that are needed to start the process do have to be harvested from an animal, but that is done without causing any harm. And the nutrient solution needed for the growth and multiplication of the cells is derived from animals, although in the future, feed based on bacteria is a possibility.

Stem cells are primitive cells that have not yet undergone differentiation, meaning that they have not yet turned into nerve cells or fat cells or muscle cells, or whatever other type of cell they are destined to become. It is possible to coax stem cells in a certain direction by nurturing them in an appropriate medium. Pig stem cells isolated from muscle tissue, for example, can be turned into muscle cells by exposing them to horse fetal serum. The cells will then multiply and produce what amounts to muscle tissue.

Dr. Mark Post at Maastricht University in the Netherlands has already produced small strips of muscle in the laboratory using this technique. That's a long way from steak, especially since the tissue, having no blood supply, looks pretty anemic.

But before long, problems will be overcome and meat will be produced in the laboratory. That will be of benefit not only to animals, but also to the environment. Raising animals leaves a large footprint in terms of greenhouse emissions and the use of energy, water and land. What the meat will taste like? Nobody knows. Issues of safety have not yet been addressed. It will be a while before sausage comes out of a test tube instead of a meat grinder.

Should you put broccoli on a pizza before or after baking?

Broccoli piled on a pizza before sliding it into the oven is tastier; broccoli added after taking it out is healthier. Life is full of difficult decisions, isn't it? But fret not. This decision can be made a little easier by examining the properties of sulphoraphane, a compound touted by a plethora of books, magazines, websites and nutritional gurus of all sorts for its anti-cancer effect. But I'll let you in on a secret: while sulphoraphane may have anti-cancer properties, it is not found in broccoli!

That doesn't mean that we are being conned about the benefits of broccoli. Far from it. There may be no sulphoraphane in broccoli, but the vegetable is still an excellent source of this compound. Sound confusing? Let's clear it up. You do have to get ready for some complicated terms though. But stick with it. It's worthwhile.

While sulphoraphane may not be present in broccoli, its precursor, glucoraphanin, is. And it is ready to liberate its anti-carcinogenic component when prompted by an enzyme known as myrosinase, also present in broccoli. In intact plant cells, myrosinase is physically separated from glucoraphanin, and only when broccoli is chewed or chopped do the two chemicals come into contact and release sulphoraphane, which can then be absorbed into the bloodstream from the small intestine. Unfortunately, myrosinase is inactivated by heat, so excessive cooking reduces the amount of sulphoraphane that can be released.

Several questions arise when considering the biological effects of broccoli. The term *anti-cancer* is commonly bandied about when talking about this and other cruciferous vegetables. How do we know it has such an effect? Actually, we really don't know. But as is so often the case in science, we can make an educated guess. Epidemiological studies show that populations that consume more cruciferous vegetables, such as broccoli, cauliflower, cabbage, kale and Brussels sprouts, have lower rates of cancer. And it is these vegetables that release isothiocyanates, a family of compounds to which sulphoraphane belongs. However, these populations also tend to consume more of many other vegetables and fruits. Furthermore, surveys based on self-reporting of food consumption are notoriously unreliable.

A few studies have examined the relationship of cancer and isothiocyanates in the urine. The idea is that amounts present in the urine reflect the consumption of cruciferous vegetables. In one such study, Chinese men who had detectable levels of isothiocyanates were at significantly lower risk of developing lung cancer over the following ten years than men with undetectable levels. Another way of looking into diet–disease relationships is the case-control study in which patients with a disease are compared with healthy controls. One such study found that urinary isothiocyanate excretion was significantly lower in Chinese women diagnosed with breast cancer than in a control group. Curiously, in the same study, cruciferous vegetable intake estimated from food-frequency questionnaires was not associated with breast cancer.

There is also some theoretical support for the benefits of sulphoraphane from laboratory cell-culture studies. When human cells are confronted with a toxin, they try to eliminate it through the action of various detoxicating enzymes. Sulphoraphane induces the formation of these enzymes, since cells regard it as a foreign substance and mark it as a target for elimination. While sulphoraphane may be harmless, some of the other compounds that are eliminated by the higher level of detoxicating enzymes may be carcinogens. Putting all this together suggests that eating broccoli regularly is a smart thing to do. The evidence is circumstantial, but there is no downside except for people who are on anticoagulant medication. In that case, eating large amounts of broccoli can be a problem because the vitamin K in the vegetable can counter the effects of the anticoagulant.

Why do cruciferous plants produce chemicals such as sulphoraphane? It isn't because they are interested in protecting humans against cancer. They are interested in protecting themselves from insects, bacteria, viruses and fungi. Attack by any of these causes cell walls to be ruptured, bringing myrosinase into contact with glucoraphanin to produce sulphoraphane, which has insecticidal and antimicrobial properties. We are the lucky beneficiaries of such clever plant activity.

Dietary supplement manufacturers are quick to jump on any study that can be used to promote the sale of products. Broccoli powder is available in health food stores, but whether or not it can provide sulphoraphane is questionable, since processing destroys myrosinase. Incidentally, the same is true for frozen broccoli. The blanching carried out prior to freezing eliminates most of the enzyme.

So, what does all of this have to do with my pizza problem? Baking the pizza with a broccoli topping is sure to destroy all the myrosinase. On the other hand, a pizza with raw broccoli is no gustatory delight. But University of Illinois researchers led by Elizabeth Jeffery have come to the rescue with their discovery of what I'll call the "spice effect." Certain spices, such as mustard, horseradish and wasabi, are rich in myrosinase, and sprinkling these onto cooked broccoli will release sulphoraphane. Now it seems we can eat our broccoli cooked and have our sulphoraphane too. All we have to do is spice up our life a little!

There's a message here about cooking broccoli as well, pizza considerations aside. Don't cook it to death! If you want to get the sulphoraphane benefits, steam it or microwave it for a couple of minutes. Remember, though, that single foods should never be looked upon either as saviours or criminals. It is the overall diet that matters. Still, eating three to five servings of broccoli a week may provide enough sulphoraphane to have an anti-cancer effect. But that's a guess. Hopefully, an educated one. And we may be able to derive even more broccoli benefits when a new variety of broccoli called Beneforté becomes available on this side of the Atlantic. British scientists used traditional techniques to crossbreed a wild variety of broccoli they discovered in Italy with regular broccoli to produce a plant that has some three times as much glucoraphanin as the currently available version.

And let me also point out that another University of Illinois study found a more significant shrinkage in prostate tumours in rats fed a combo of tomato concentrate and broccoli than with either food alone. So, a broccoli pizza with lots of tomato sauce sounds better and better. Even if you are not a rat. Don't forget to sprinkle on the fresh horseradish before you dig in.

MIRROR, MIRROR ON THE WALL

What are cosmeceuticals?

Think of cosmeceuticals as drugs that are designed to improve appearance from the inside out. Strength Within, first introduced in the U.K. in 2011 is an interesting case in point. When these capsules, the product of at least five years of research by a team of scientists at Unilever's laboratories, first appeared on the market, they sold out after just four hours. Pop some, went the promise, and wrinkles, those fearsome hallmarks of aging, would be ironed out from the inside!

When word leaked out that the beauty pills would be test-marketed for two weeks, they were scooped up like candy. After all, no prescription was needed. The ingredients—soy isoflavones, omega-3 fats, lycopene, vitamin E and vitamin C—all occur naturally in food and can be marketed as supplements. No further approval is necessary unless there are claims of treating, preventing or curing a disease. Wrinkles may terrorize, but they can hardly be regarded as a "disease." So treating them is not a medical claim, hence Strength Within is regarded as a cosmetic, not a drug. *Cosmeceutical* describes such products that blur the line between cosmetics and drugs.

Improving the skin's appearance from within is not a novel idea. After all, acne medications do exactly that. And a plethora of dietary regimens have claimed to reveal the secret of achieving beauty by the bite. Many of these also take a big bite out of the pocketbook, with Dr. Nicholas Perricone's "wrinkle cure" being a prime example. Perricone is a dermatologist with frequent TV appearances in which he outlines his formula for reversing wrinkles and sagging skin. He claims that increasing protein intake, while avoiding foods like bananas, bread, rice, beets and sweet potatoes, is the key to healthy skin. Perricone is high on salmon, which he calls "your magic bullet for great skin tone, keeping your face firm and contoured." My face changes its contour when I hear stuff like that, because there is just no evidence for such assertions. And his costly dietary supplements and creams may be "designed to reduce the appearance of loss of tone, sagging skin, and fine lines," but there is no peer-reviewed evidence to show that they actually do. The Perricone regimen will, however, reduce one's bank account to the tune of over $400 a month. Not only in appearance, but in substance.

None of this is to suggest that diet doesn't play a role in the condition of the skin. Indeed, it plays some role in virtually everything that goes on inside of us, since food is the only raw material that ever enters our body. Rosacea, for example, can often be triggered by alcohol or spicy foods, lack of niacin can cause dermatitis, and acne can be exacerbated by specific foods, depending on the individual.

Furthermore, a balanced diet can have a major impact on how skin ages as well as on the risk of skin cancer. The biggest factor in skin aging is the damage carried out by free radicals generated by exposure to sunlight. Vitamins C and E, along with beta carotene, can sop up the nasty free radicals, and vitamin C also plays a role in building new collagen, the protein that gives skin its resilience. A diet that includes lots of fruits and vegetables, especially beta carotene—rich ones like sweet potatoes (one of Perricone's supposed nemeses), while reducing fats is the way to go. A Baylor University study effectively demonstrated that patients who had been treated for skin cancer were able to significantly reduce the chance of recurrence by maintaining a diet in which only 20 per cent of the calories derive from fat, as opposed to the North American average of 38 per cent.

Being told that the route to healthy skin involves eating a balanced diet, shunning smoking, getting plenty of exercise and staying out of the sun is not what some people want to hear. Swallowing a few Strength Within capsules seems a more attractive proposition. This, however, is not the first product to offer hope in a bottle. Imedeen tablets, developed by the Danish healthcare company Ferrosan (recently acquired by Pfizer), have been around since 1991. And unlike most cosmeceuticals, they have placebo-controlled randomized trials to back up the claims of improved skin appearance. A mix of soy extract, fish polysaccharides, white tea extract, grape seed extract, tomato extract, vitamins C and E and zinc, Imedeen was tested in eighty post-menopausal women. Observations by independent evaluators, as well as objective data from ultrasound measurements, backed the claims of improvement when compared with placebo.

But, there is a 'but.' It took six months for Imedeen to work, and the effects could hardly be described as spectacular, though they were at least statistically significant. However, there is a difference between statistical significance and practical significance. Especially when it comes to forking out $500 for a six-month supply.

Unilever's Strength Within is very similar in composition to Imedeen, but it adds a new wrinkle of its own. Researchers managed to show that this particular chemical mix activated genes responsible for collagen synthesis, and even confirmed through biopsies that new collagen had formed in the dermis, the skin's deepest layer, one that is generally not reachable by creams. Of course, the real question is whether this "gene food" treatment will elicit "gee, you look great" type of comments. Judging by Unilever's double-blind trial, I would put that in the doubtful category.

After fourteen weeks, the crow's-feet wrinkles around the eyes became on average 10 per cent shallower in subjects treated with Strength Within. Frankly, that is not a very gripping difference. But it is proof of concept. Maybe improved versions of the capsules will produce improved effects. It would also be interesting to see trials comparing oral supplements to the effects of topical treatments with moisturizing creams, some of which have also been shown to help reduce wrinkles. All in all, though, it is a person's inner beauty that is attractive, and that cannot be obtained from a pill.

The activist organization "Campaign for Safe Cosmetics" garnered a great deal of publicity with the release of a report entitled "Baby's Tub Is Still Toxic." What chemical were they up in arms about?

The warning was all about the trace amounts of formaldehyde or formaldehyde-releasing compounds added to some baby shampoos to prevent bacterial contamination. Reading the report might send any mother who has used a baby shampoo preserved with formaldehyde into a state of panic, fearing they might have doomed their offspring by exposing them to a "known carcinogen." Well, let's throw in a little science before we throw out the baby shampoo. Yes, formaldehyde is a likely carcinogen. But that categorization was made on the basis of inhaling the chemical in significant amounts during occupational exposure. Embalmers and pathologists have indeed experienced a slight increase in cancer rates attributed to formaldehyde, mostly of the nasal cavity. However, data indicate that no tumours have been found when occupational exposure was below 2.4 milligrams of formaldehyde per cubic metre of air.

Now let's crunch a few numbers. A shampoo preserved with a formaldehyde-releasing agent such as quaternium-15 has a formaldehyde-yielding potential of 0.4 milligrams per gram. And let's say that about 10 grams of shampoo—a rather generous amount—are used to wash baby's hair. So we have the possibility of releasing 4 milligrams of formaldehyde into the air. But wait! Formaldehyde is very soluble in water, so very little will evaporate. But let's really play it safe and assume that half, or 2 milligrams, will evaporate.

Now, assuming that the volume of air in a bathroom is about 10 cubic metres, and that there is zero ventilation, the concentration of formaldehyde will be 0.2 milligrams per cubic metre, or one-tenth the amount that has never caused a tumour, even with continuous exposure. Of course, here we are not talking about continuous exposure. Nobody bathes a baby for eight hours a day. Therefore, any suggestion that formalde-hyde in shampoo presents a cancer risk is unfounded. Actually, Mommy's breath may be a greater risk. Formaldehyde is a product of human metabolism, and breath can contain concentrations of formaldehyde at 0.4 milligrams per cubic metre. Maybe the greatest benefit of fearmongering about formaldehyde in shampoo is to make Mommy breathless.

While inhalation of formaldehyde from shampoo is a non-issue, problems caused by skin exposure cannot be so readily dismissed. Formaldehyde is a known allergen and can trigger rashes and inflammation, but these effects are also dose dependent. Skin sensitization can occur when the concentration in a solution is above 0.2 per cent. And how much is present in baby shampoo? If all the formaldehyde were instantly released from quaternium-15, the concentration in baby shampoo would be 0.04 per cent. But quaternium-15 is a slow releaser, so the effective concentration is far less than that. There may be some babies that react to these remarkably small amounts, but far fewer than would react to the bacterial contaminants in a poorly preserved product.

As can be seen, once appropriate calculations are made, the clamour to remove formaldehyde as a preservative from baby shampoos amounts to no more than unscientific noise. But as far as marketing goes, the public is right even when it is wrong, and manufacturers are working towards using alternate preservatives such as benzyl alcohol, potassium sorbate, sodium benzoate or methylisothiazolinone. Of course one can dredge the scientific literature and come up with risks for these as well. Dig deep enough, and at some dose you can find some sort of risk with any chemical.

Furthermore, if we want to eliminate all risks from formaldehyde, we'll have to get rid of particleboard, permanent-press fabrics, varnishes, paints, carpeting, curtains and nail polish, as well as many types of insulation and paper products, all of which are manufactured with formaldehyde and can outgas the compound. In some cases, as with the notorious trailers that were supplied to victims of the Katrina disaster, outgassing can be enough to cause severe eye irritation and respiratory symptoms. To curb formaldehyde exposure, we would also have to forget about fireplaces, gas cookers, driving cars and, of course, smoking. Even then, we would be ingesting some formaldehyde because it occurs naturally in virtually all foods.

When it comes to the use of products for the hair, there is one area where formaldehyde poses a real concern. Some of the "Brazilian" hair-smoothing products contain sufficient formaldehyde to pose an occupational hazard to hairdressers. Here, the chemical is not used as a preservative, it is one of the active ingredients needed to form links between protein molecules in hair. And unfortunately, the presence of formaldehyde in these products can be hidden by using alternate names for the chemical on the label. Methylene glycol, methanal, oxomethane and formalin are some of these. Often, these products declare themselves to be "formaldehyde-free."

You certainly wouldn't want a baby lying all day in a hairdressing salon where such products are used to "smoothen" hair. But washing baby's hair with baby shampoo, any kind of baby shampoo, is a completely different story. So, let's get formaldehyde out of Brazilian hair products, but let's not throw out the baby with the bath water. Trace amounts of formaldehyde in shampoos prevent bacterial contamination, a significant problem, at minimal risk.

If you want to worry about a real risk in the tub, consider that about 150 children in North America, the majority under four years old, end up in emergency rooms each year mostly because of tub falls, many of which could be prevented with the use of a vinyl mat. Now, that really matters. But I suspect that the folks behind the Campaign for Safe Cosmetics might then make an issue of the phthalate plasticizers that leach out of the mat.

What popular soap owes its name to the fact that it is made from oil extracted from the fruits of two tropical trees?

Palmolive soap is made from a mixture of palm and olive oils. All soaps are made by reacting fats with sodium hydroxide, commonly known as lye. Any source of fat will do, be it beef tallow or vegetable oil. Soaps were made from animal fat until about the sixteenth century, when Europeans discovered that vegetable oils could be used to make a product that was less harsh on the skin.

At the end of the nineteenth century, B.J. Johnson, an American soap manufacturer, was producing a soap from palm and olive oils that proved to be so popular that Johnson renamed his company Palmolive. In 1928, Palmolive bought out a rival soap manufacturer and became the Colgate-Palmolive Company. Colgate had been around for almost a hundred years and had been profitable enough to donate large funds to Madison University, which was eventually renamed Colgate University.

Today, Colgate-Palmolive is a huge multinational corporation that produces a variety of health care and personal products, often in fierce competition with Procter and Gamble, the world's largest soap and detergent manufacturer. Both companies have advertised their products widely, bringing fame to "soap operas." Back in the 1920s, Palmolive's ads focused on the soap's supposed ability to keep complexion fresh and smooth by keeping the skin scrupulously clean. One ad suggested that "the blushing bride of today should be the blooming matron of tomorrow, retaining the charm of girlhood's freshness to enhance radiant beauty." "With a fresh smooth skin," the ad went on to say, "no woman ever seems old." And, of course, Palmolive was the soap that "freshens and stimulates, encouraging firmness and attractive natural colour."

Today's ads deliver a similar message, encouraging young ladies to try Palmolive and "feel the difference in just seven days." Whether this claim can be scientifically supported is a matter of controversy, but this controversy is trivial compared with two others that surround palm oil.

Palm oil is the world's most widely produced vegetable oil, extracted from the flesh of the fruit of the palm oil tree, grown on plantations in Southeast Asia, mostly in Malaysia and Indonesia. The oil has far more applications than soap manufacture. It is used to make a variety of cosmetics, to produce biodiesel and, most importantly, in the production of foods such as margarine, shortening, baked goods, microwave popcorn and candies. The controversies surrounding palm oil swirl around its environmental consequences and its effects on health. "Orangutans are literally dying for cookies," claims one environmental organization. What's the connection? Rain forests, the orangutan's habitat, are being cut down to expand palm oil tree plantations.

Greenpeace has drawn attention to this situation using a variety of disturbing ads, one of which targets the Nestlé company and one of its products, Kit Kat. The ad shows a man opening and unwittingly biting into a Kit Kat that contains an orangutan finger, with bloody results. The point is to urge Nestlé to stop buying palm oil to produce its products.

The palm oil industry doesn't take this lying down; it fights back with arguments and statistics demonstrating that palm oil can be produced in an environmentally sustainable fashion without endangering wildlife. It claims that Greenpeace endangers human life by taking away the livelihood of people who work on plantations. The truth is difficult to decipher—as it is with the nutritional controversy.

Palm oil contains about 50 per cent saturated fat, the type of fat linked with heart disease. Its use has increased with the campaign to eliminate trans-fats from the diet because palm oil has similar properties to the hydrogenated vegetable fats that harbour the notorious trans fats. Palm oil is stable to heat and is semi-solid at room temperature. Anti–palm oil activists maintain that all saturated fats are bad, while the palm oil industry claims that palmitic acid, the saturated fat in palm oil, does not raise cholesterol in the same way as saturated fats in dairy and meat products. Both sides muster studies to prove their point, and once again the consumer is left confused. Scientists, too. Me included.

✧

What personal-care product owes its effectiveness to the breakdown of an emulsion and the consequent release of dimethicone?

A two-in-one shampoo. Talk about a chemical triumph! You can now clean and condition your hair at the same time.

So, what's the big deal, you ask? The development of a two-in-one shampoo was a great challenge because such a product requires leaving a conditioning agent behind just as a dirty oil layer is removed. That oily layer is sebum, a secretion of the sebaceous glands that surround the follicle from which hair grows. Sebum prevents hair from drying out but is also a magnet for dirt. The role of a shampoo is to strip away the grimy sebum. This is not a difficult task, and is readily accomplished by a *surfactant.* Surfactant molecules have an oil-soluble end and a water-soluble end, allowing for the formation of links between the soiled oily layer and the rinsing water. Soap is the simplest surfactant, but it does present a problem if the water is hard, meaning it contains dissolved calcium and magnesium. These minerals react with soap to form a "scum" that deposits on hair. Not a good thing.

The answer to this problem is a different kind of surfactant, known as a detergent. Detergents work just like soap but do not react with minerals to form a precipitate. But there is another issue with surfactants. As the hair's protective layer of sebum is stripped away, it is left feeling dry and "flyaway"—in other words, it is in poor condition.

Enter the conditioners. These are chemicals that are applied after shampooing to coat and protect the hair in a fashion similar to sebum. Dimethicone is a synthetic silicone oil that coats the hair fibres, leaving them smooth and shiny. It is a common ingredient in conditioners. But how can it be incorporated into a shampoo so that it stays behind instead of being washed away like the sebum it is attempting to replace? This is where clever chemistry comes in. It all has to do with formulating an appropriate emulsion.

An emulsion is a mixture of two or more immiscible liquids in which one of the liquids is dispersed in the other. Milk is a common example, with tiny globules of fat being suspended in an aqueous phase. Salad dressings are also emulsions. Shaking a mixture of oil and vinegar produces an emulsion, but unless this emulsion is stabilized somehow, it will break down and separate into two layers. This is where an *emulsifier* comes into the picture. Emulsifiers are molecules that stabilize an emulsion by positioning themselves at the interface of the two phases, preventing the suspended droplets from coalescing. In the case of a vinaigrette, a bit of lecithin found in egg yolk will do the job.

Forming an emulsion is not that difficult. Cosmetic chemists have plenty of experience in this area since virtually all creams and lotions are emulsions. But designing an emulsion that breaks down at an appropriate time is quite a task. Yet that is exactly what is needed for a two-in-one shampoo. The key is to keep the conditioning agent, usually dimethicone, suspended in an emulsion while the detergent does its job of removing sebum, and then have the emulsion break down as copious amounts of rinsing water are applied. It all depends on finding the right emulsifier.

After much experimentation, dehydrogenated tallow phthalic acid amide turned out to be just right for the job. It keeps the tiny droplets of dimethicone in suspension while the hair is being washed, and allows the emulsion to break down when the hair is flooded with water. The dimethicone then is released and deposits on the hair. Smooth and silky, here we come, two jobs have been done in one.

☆

In what sort of personal-care product would you find "storax"?

To be honest, I had never heard of storax until the word caught my ear while watching *Perfume: The Story of a Murderer*. Based on Patrick Süskind's 1985 novel, it's a fascinating film, especially for anyone interested in chemistry. Jean-Baptiste Grenouille, a man who has no body odour himself, becomes obsessed with smells and dreams of creating the "perfect" perfume. Believing the ideal ingredient to be the scent of young virgins, Grenouille embarks on a murderous spree to collect the prized ingredient. Using a perfumer's technique commonly applied to botanical matter known as *cold enfleurage*, he wraps the murdered women in linen soaked in animal fat. Their fragrance diffuses into the fat, from where he proceeds to extract it with alcohol.

The wretched man had learned the art of perfumery from master Giuseppe Baldini, who had hired him after a spectacular impromptu scent performance. Baldini had been trying to reproduce Amor and Psyche, a rival perfumer's popular product, when Grenouille happened to wander into his shop. After sniffing the original, Grenouille immediately concluded that it contained cloves, roses and storax! He proceeded to revitalize Baldini's fading career by creating a host of novel smells, but was continually haunted by the desire to formulate the "perfect" fragrance.

I knew that roses and cloves were common ingredients in perfumes, but what was storax? The search for an answer took me on a journey from ancient tropical trees to modern computer housings. Storax, it turns out, is a resin produced by a number of tropical trees of the family *Styracaceae* when their bark is injured. It has a long history of use in perfumes because of its lingering fragrance and its ability to slow the evaporation of other compounds that contribute to the overall scent. This *fixative* effect allows a perfume to keep the original fragrance for a longer time. Like other fixatives, such as sandalwood, patchouli, frankincense, benzoin, cedarwood, ambergris, musk and castor oil, storax is a complex mixture of compounds. Cinnamic acid, alpha-pinene, ethyl cinnamate and vanillin are just some that have been isolated. But in terms of historical impact, perhaps the most interesting compound found in storax is styrene.

In 1839, Eduard Simon, a German apothecary, was attempting to separate, by distillation, the components of storax obtained from the *Liquiambar orientalis* tree. One of the fractions he collected was an oily substance that seemed to be a single compound. Simon named it "styrol" and stored it in a bottle. Within a few days, and much to his surprise, his styrol had changed from an oil into a hard, translucent mass. Since he hadn't added anything to the sample, Simon figured that it must have reacted with oxygen, and dubbed the new material "styrol oxide." As it turned out, he was wrong. August Wilhelm von Hofmann, one of the leading lights of German chemistry, showed that the styrol transformation also occurred in the absence of oxygen.

A solution to the mystery was proposed in 1866 by the brilliant French chemist Marcellin Berthelot. The molecules of styrol, Berthelot suggested, must have joined together to form a long chain. Just three years earlier, he had delivered a landmark lecture to the Chemical Society of Paris in which he introduced the novel idea of small molecules linking together to form giant molecules, or polymers. But Berthelot did more than theorize. He carried out experiments to show that ethylene molecules could be joined together to form a new substance he dubbed "polyethylene," surely the first time that term was ever used. The history of polymer chemistry can be said to have begun with Berthelot's pioneering work.

Simon's styrol was eventually identified as the compound we now know as styrene, and his styrol oxide was actually polystyrene. Neither Simon nor Berthelot found a practical use for polystyrene, but by the 1930s German chemists did. They discovered that it could be cast into virtually any shape by pouring the molten substance into moulds. It could also be extruded into sheets or films. Today, polystyrene is used to make myriad items ranging from clear plastic glasses, laboratory equipment and jewel cases for compact discs, to smoke detectors and disposable razors.

The most common use for polystyrene, however, is to produce *expanded polystyrene.* That's the foamy stuff of coffee cups, insulation materials and those packing peanuts used to cushion electronic equipment. The "expansion" is produced by blowing pentane or carbon dioxide into the melted plastic. Pentane is a liquid, but it quickly evaporates to yield a gas, and carbon dioxide, of course, is a gas. As these gases attempt to escape from the molten polystyrene, they get trapped as the material cools, forming pockets of gas. The result is polystyrene foam. Styrofoam is the Dow Chemical Company's trade name for its version of foamed polystyrene. Of course, the styrene needed to make polystyrene is no longer obtained from the distillation of storax. It is now produced on a gigantic scale from benzene and ethylene derived from petroleum.

While polystyrene is a very useful substance, it is overused. We need computer housings and insulation material, but do we need all those disposable razors, cutlery and coffee cups? Polystyrene can be recycled, but it is not always economical to do so, and it often ends up in landfills or being incinerated. That needs to change. We have to conserve our petroleum resources and our environment.

What are soap nuts?

No, they are not people obsessed with cleaning. Nor are they any kind of nut. Soap nuts are actually berries that grow on *Sapindus mukorrosi* trees native to India and Nepal, and they derive their name from the fact that they can be used for cleaning purposes like, well, soap. The berries are blessed with a significant content of saponins, compounds that have soaplike cleaning activity, although chemically they are very different from soap. Saponins do, however, share an important feature with soap. Both are molecules that have distinct water- and fat-soluble regions. The fat-soluble part embeds in greasy dirt, while the water-soluble part bonds to water. The result is that grease gets rinsed away with water.

Like soap, saponins also congregate at the surface of water droplets and act to reduce the attraction between water molecules. This reduction in *surface tension* allows water to spread over a surface more readily and leads to easier penetration of fabrics. Soap berries have a long history of use as a cleaning agent thanks to their high content of saporins, but other plants also contain these compounds.

Saponins actually constitute a large family of compounds with two common properties: they all produce foam when shaken with water, and they all have a basic molecular structure composed of a carbohydrate part and a complex network of carbon atoms arranged in rings. Some saponins have interesting physiological activity. Digoxin, the heart stimulant found in foxglove, is a saponin, as is solanine, the bitter substance in green potatoes. Some saponins found in tree barks have even been linked with a cholesterol-lowering effect.

The nomadic Masai of Kenya live almost exclusively on meat and milk and consume upwards of 2 grams of cholesterol a day, a stunning amount. Yet they have a remarkably low incidence of heart disease and a healthy blood cholesterol level of 3.5 millimoles per litre. Some researchers suggest that the secret lies in the saponin content of tree bark that Masai regularly add to their food. Saponins, they say, can prevent the absorption of cholesterol. There are other folkloric uses of saponins, particularly those found in soap nuts, as well. Supposedly, they can help with migraines, inflammation and infections. Even claims of contraceptive properties have been put forward. Maybe if the nuts are held between the knees.

While any such health claims of dietary supplements featuring saponins should be viewed skeptically, the cleaning ability of these compounds has been well demonstrated. The berries are dried, and a handful are put into a little cotton bag that is placed in the washing machine with the laundry. After the laundry has been washed, the berries can be dried and reused several times. In this era of greenness, soap nuts are often promoted as an environmentally friendly alternative to detergents.

And in this case, there is something to the argument. Soap nuts are a renewable resource, their production does not rely on using fossil fuels, they don't contain any added fragrances that can cause allergic reactions, saponins are biodegradable, and soap nut shells are compostable.

Of course, the important question to ask is whether they clean as well as commercial detergents. No, they don't. They won't get out stubborn stains as well as scientifically formulated laundry products, with their array of surfactants, water softeners, enzymes and bleaching agents, but for your average load of laundry, they'll do an adequate job.

Soap nuts are correctly advertised as being less irritating to the skin than other commercial laundry products. But they do have one irritating feature: they are often promoted as "chemical-free." That, of course, is ridiculous. What are saponins if not chemicals?

WEIRD
AND
WONDERFUL

What happens to crystals of Epsom salt if you yell at them?

They turn into a powder! It isn't fear that causes the crystals to crumble, it is loss of moisture. Epsom salt, so called because it was first mined at Epsom in the U.K., goes by the chemical name magnesium sulphate heptahydrate. This means that the crystal structure incorporates seven molecules of water for each unit of magnesium sulphate. If one of these water molecules is lost, the resulting hexahydrate doesn't maintain the crystal structure, and the Epsom salt crumbles into a powder. The loss of a water molecule requires heat. And where does the heat come from? Surprisingly, from the absorption of sound.

Magnesium compounds have the interesting property of absorbing sound waves and converting the sound to heat. This property is partially responsible for an urban legend that claims placing a dish of Epsom salt near a microwave oven reduces microwave emissions. Supposedly, the crumbling of the crystals means that any leaking radiation has been absorbed by the crystals, preventing the radiation from reaching people who might be around. This is absolute nonsense. True, leaking microwaves could heat up the water that is incorporated in the crystals and cause it to evaporate and turn the crystals into a powder. But suggesting that this offers protection from radiation is like suggesting that a sunbather beside you is absorbing the sun's rays, thereby protecting you from being burned. In any case, modern microwave ovens do not leak, so the crumbling of the Epsom salt is probably coming from sound absorption.

There is a more serious side to the ability of magnesium sulphate to absorb sound. That story takes us to the ocean. Sound waves travel very well through water; in fact, they travel faster than through air. Of course, transmission falls off with distance, but it is also curbed by chemicals dissolved in the water. Magnesium sulphate and boric acid are the two main substances in ocean water that absorb sound, and interestingly, their efficiency at doing so depends on acidity of the water. Magnesium sulphate is composed of positive magnesium ions and negative sulphate ions, and the closer the ions are to each other in solution, the better the sound absorption. As the acidity of the water increases, the attraction between the ions decreases and sound transmission improves.

What is the point of all of this? The acidity of the oceans is increasing as we spew out more and more carbon dioxide. Some of the gas dissolves in the oceans, forming carbonic acid. Experiments show that even a tiny increase in acidity results in better sound transmission, which in turn can have all sorts of effects on ocean life. Human-generated noise from the surface can scare fish away more readily, and the background noise can also interfere with the behaviour of marine animals that communicate with each other through sound. Nobody knows what the exact consequences may be, but it is clear that carbon dioxide emissions may have far more reaching effects than causing climate change.

Why did the Lone Ranger's silver bullets not tarnish?

Because the concentration of hydrogen sulphide in the desert air of the Old West was undoubtedly very low. Contrary to popular opinion, the tarnish on silver is not caused by reaction of the metal with oxygen in the air. The culprit is hydrogen sulphide, a gas that reacts with silver to form solid silver sulphide. Hydrogen sulphide occurs in air both naturally and as a result of human activity. Many proteins contain sulphur and release hydrogen sulphide when they decompose. The odour of rotten eggs, for example, is due to this compound, and natural gas, which is the end product of the decomposition of organic matter, is also a rich source of hydrogen sulphide.

Volcanic eruptions and natural hot springs also release the gas. And it forms as well when carbonyl sulphide, a sulphur compound emitted from volcanoes and deep-sea vents, reacts with moisture. Then there is human activity. Petroleum contains a variety of sulphur compounds that have to be removed in order to prevent the formation of sulphur dioxide when petroleum burns. In the *hydrodesulphurization* process, petroleum is reacted with hydrogen, resulting in the conversion of sulphur compounds to hydrogen sulphide. The hydrogen sulphide is then converted to sulphur, but inevitably some of the gas is released into the atmosphere.

Since there were no petroleum refineries in the Old West, nor sea vents or volcanoes, tarnishing of silver was not likely to be extensive. Unless, of course, the Lone Ranger and Tonto were fond of beans. Human flatus is a rich source of hydrogen sulphide. But there is no evidence that the duo favoured beans, so we can assume that the Lone Ranger's silver bullets were bright and shiny. Where the masked man got those bullets is more of a mystery. He surely didn't melt silver over a campfire and pour it into moulds. Lead bullets can be made that way because lead has a relatively low melting point of 327 degrees Celsius. But silver melts at a scorching 960 degrees Celsius, and you don't get that over a campfire. You also need a graphite mould to stand up to that temperature.

Silver bullets can be made, but it isn't easy. And it isn't necessary, either. Lead bullets do a very good job. Unless, of course, you have to confront a werewolf. Then you had better know where to find a silver ingot, a gas furnace and a graphite mould. To keep your bullets nice and shiny if you are not in a desert, you'll need to coat them with a cellulose nitrate lacquer, better known as clear nail polish, or wrap them in a silver nitrate–impregnated cloth. The silver nitrate reacts with the hydrogen sulphide in the air preferentially, leaving the silver bullet untarnished, ready for any lurking werewolves.

What toy was banned in New York City in 1922 because of a concern over flammability?

The hydrogen-filled balloon. How can you have a proper party without balloons? Well, you can't. Today's lighter-than-air toy rubber balloons are filled with helium, but that gas wasn't available when rubber balloons were first made by Michael Faraday in 1824. Faraday was one of the greatest scientists in history and is perhaps best known for his work in electricity and magnetism as well as for his wonderful Christmas lectures to the public at the Royal Institution in London. The properties of gases were featured prominently in Faraday's presentations, but their storage presented problems. The only suitable containers were fashioned out of the bladders, stomachs or intestines of animals, mostly pigs, but these leaked easily and were smelly.

This technology goes back to the ancient Aztecs, who shaped the inflated bladders of cats to look like dogs or donkeys to be used as a sacrifice to the sun. The balloon animals were carried to the top of the Aztec pyramid and burned, supposedly to please the sun god. Apparently, when a disease wiped out cats, the Aztecs turned to human sacrifices for their supply of bladders.

Faraday had used pig bladders to store laughing gas (nitrous oxide), the properties of which he often demonstrated to the great delight of audiences. But the bladders could not keep the gas for a long time, so he began to experiment with other materials. Rubber, produced by air-drying the latex of the rubber tree, was already known, and Faraday had the idea of cutting two round pieces of rubber and coating them with flour except around the edges. The flour prevented the rest of the rubber from sticking together while he used heat to seal the edges. When filled with hydrogen, these rubber balloons became an impressive addition to Faraday's lectures, especially pleasing children, who were taken by the gravity-defying effect. It didn't take long for balloons to be commercialized—in fact, only one year.

By 1825, Thomas Hancock had devised a balloon-making kit to be sold as a toy. A bottle of rubber solution packaged with a syringe for inflation allowed children to make their own balloons. But balloons really took off when Charles Goodyear discovered that heating latex with sulphur yielded vulcanized rubber, which was more stretchy and stood up better to changes in temperature.

By the turn of the twentieth century, hydrogen-filled balloons had become a staple at parties and celebrations. They had also become a concern for firefighters, since a spark could easily set them off. And, of course, when boys discovered that—well, boys will be boys. When one of them exploded a hydrogen balloon as a prank at a city function, badly burning an official, the city took action. No more hydrogen balloons! No great disappointment for children, though save for rambunctious boys, since helium balloons were already becoming available. Indeed, by 1928 the classic Macy's Thanksgiving Day parade featured a variety of helium-filled balloons.

Both the parade and the balloons have expanded since, with the giant helium-filled balloons, now made of polyurethane instead of rubber, delighting the two and a half million spectators who line the streets and the 44 million who watch on television. A long way from inflated pig bladders.

You coat a metal wire with a mixture of potassium perchlorate, powdered aluminum and dextrin. What have you made?

A sparkler. The brilliant shower of sparks that characterizes a burning sparkler consists of glowing metal particles, usually aluminum, but iron titanium, zinc or magnesium can also used. They glow because they are undergoing combustion. In other words, the little metallic particles are burning. The metal is undergoing reaction with oxygen to form metal oxides, but in order to produce the intense glow, more oxygen is needed than is available from the surrounding air. This is where the potassium perchlorate comes in. It is an oxidizing agent, meaning that it is a source of oxygen. Potassium chlorate undergoes a chemical reaction, initiated by the lighting of the sparkler, whereby it decomposes to yield potassium chloride and oxygen. The oxygen then combines with the metal, allowing it to burn. This combustion process produces heat, which then decomposes more potassium perchlorate, yielding more oxygen, which then combines with more metal, and on and on until one of the reagents runs out.

Dextrin is a carbohydrate, composed of short chains of glucose molecules, produced by the partial breakdown of starch. Like starch, it can act as a glue when moistened, allowing the metal particles and the potassium perchlorate to bind onto the wire. It also burns during the process, producing heat to help decompose the perchlorate and thereby enhance the combustion of the metal. The sparkler's sparks do not travel far, since the metal particles are very small and burn up quickly. But touching the sparkler while it is spewing its glowing particles can lead to a nasty burn. Burning aluminum can do more than brighten up a birthday party.

The solid fuel boosters of the space shuttle were essentially giant sparklers. They were filled with a mixture of aluminum powder and ammonium perchlorate held together with a rubber binder. The intense glow emanating from the booster at liftoff was caused by the extremely exotheric reaction of aluminum with the oxygen provided by the decomposing perchlorate. By Newton's third law—for every action, there is an equal but opposite reaction—the shuttle blasted upward as the hot combustion products emerged from the tails of the boosters. Unfortunately, we have seen the last of these giant sparklers. All the space shuttles have been retired.

What field of science is based upon Locard's Principle?

Forensic science. In its simplest form, Locard's Principle states that when there is contact between two items, there will be an exchange of something. That something can be dirt, sweat, fibres, hair, saliva, paint, cosmetics, semen, blood, skin, footprints or fingerprints. In other words, a criminal always takes something away from the scene of a crime and leaves something behind. Today, crime scene investigators use a variety of sophisticated techniques to detect trace materials that may have been left by a criminal, but the world's first lab dedicated to scientific crime investigation was only established in 1910 by French medical examiner Edmond Locard. "It is impossible for a criminal to act without leaving traces of his presence," was Locard's thesis, ably demonstrated by his clever solving of the murder of Marie Latelle in 1912.

Suspicion fell on Marie's boyfriend, Emile Gourbin, who professed innocence and claimed to have been playing cards with his friends at the time of the murder, a story they corroborated. But when Locard examined scrapings from Gourbin's fingernails under a microscope, he found traces of a pink powder that appeared to be makeup. Locard managed to locate a chemist who specialized in custom-made powders and discovered that he had indeed formulated one for the murder victim. Microscopic examination confirmed a match, and when confronted with the evidence, Gourbin confessed. He also described how he had tricked his friends into offering an alibi by changing the time on the clock in the room where they played cards.

Although Locard was the first to establish a lab devoted to criminal investigation, it was Sir Arthur Conan Doyle who laid the foundations to forensic science through the adventures of Sherlock Holmes. In a 1904 story, "The Adventure of Black Peter," Holmes remarked that "as long as a criminal remains upon two legs so long must there be some indentation, some abrasion, some trifling displacement which can be detected by the scientific researcher." A truly prophetic statement.

Today, the tiniest of paint chips can be used to identify the make and year of a car, solubility tests can readily distinguish between different types of synthetic fibres, and even different types of nylon fibres can be identified by their melting points. DNA isolated from trace amounts of body fluids can be linked to specific individuals. Locard was right. Every contact leaves a trace.

In 1861, a cartoon depicting whales rejoicing at a party appeared in *Vanity Fair*. What were the whales celebrating?

The whales were thrilled about the discovery of oil in Pennsylvania. At the time, whales were mercilessly hunted for their oil, which was the ideal fuel for lamps. But much to their relief, in 1859, Colonel Edwin Drake drilled his famous well near Titusville and the hunt for oil shifted from the waters of the oceans to the soil of Pennsylvania. Oil had long been known in the area, but was usually disdained for contaminating salt wells. In the 1800s, extracting salt from underground deposits was a profitable industry in Pennsylvania. A well would be drilled and water piped down to dissolve the salt. Then the salty water would be pumped to the surface and allowed to evaporate, leaving crystalline salt behind.

But sometimes, much to the annoyance of the salt miners, the salty water coming out of a well was contaminated by oil. They didn't know what to do with the cruddy material, but Samuel Kier had an idea: sell it as medicine! Kier's Rock Oil claimed to cure burns, ulcers, cholera, asthma, indigestion, rheumatism and even blindness. Business must not have been too great, possibly because of the blindness claim. That claim is not usually made by the producers of quack products, because, let's face it, a blind person can readily determine if the product works or not. In any case, Kier, looking for other uses for oil, sent a sample off to a chemist in Philadelphia, who suggested distilling it to collect a fraction that could be sold as "carbon oil" to be used in lamps.

Kerosene, as we now call the distillate, became popular and stimulated Kier to build the first-ever petroleum still. Others quickly got into the game, and the hastily formed Seneca Oil Company hired Colonel Edwin Drake to explore the possibility of drilling for oil. Drake was not a colonel of any kind, but the company thought a military title would be more attractive to investors. Drake in turn hired William Smith, an experienced salt well digger. Things did not go well, and the investors were losing patience.

But on August 28, 1859, Smith noticed oil floating in a hole they had been drilling, and the Pennsylvania oil boom was under way. The very same day, Drake received a letter from the investors, asking him to forget the whole business and close up shop. Of course, the opposite happened, and almost overnight Titusville grew from 250 residents to over 10,000.

Jonathan Watson, the gentleman who owned the land where Drake had drilled his well, became the world's first oil millionaire. Until the Texas oil boom some fifty years later, Pennsylvania produced half the world's oil. Unfortunately, it also was the site of the first major oil accidents. On Black Friday, as it came to be known, in 1880, a lightning strike ignited 300,000 barrels of oil, causing a fire that burned for three days. Another lightning strike fourteen years later caused an explosion that killed sixty people.

Drake's discovery changed the world, in more ways than one. Oil would eventually heat our homes, fuel our cars and source the myriad chemicals that define modern life. But it would also set the stage for climate change.

※

What is a "kangatarian"?

If the only meat you eat is kangaroo, you are a kangatarian. And kangatarians are multiplying, although not as quickly as the marsupials they feast on. What is the appeal of eating "roo"? It isn't the taste. The meat is gamey and can be very dry. Most kangatarians are driven by environmental and ethical issues, while some are attracted by the healthier fat profile of the meat. Indeed, kangaroo steak has far less fat than beef or lamb cuts, and even less than lean chicken breast. Even better, hardly any of the fat is the "saturated" variety. Rather, kangaroo meat has a high concentration of conjugated linoleic acid (CLA), a fat that has been linked with reduced risk of obesity, cancer, diabetes and atherosclerosis.

But the majority of kangatarians are not interested in the health benefits, which in the context of an overall diet are probably minor. They are interested in the health of the environment as well as in the humane treatment of animals. How can eating kangaroo help the environment? It has to do mostly with the microbes kangaroos have in their digestive tracts. Unlike those found in the rumen of cattle or sheep, these microbes produce virtually no methane as they help digest the animal's food. Methane is a potent greenhouse gas, and calculations show that replacing just one-fifth of beef by kangaroo meat in Australia would lead to a 28 per cent cut in greenhouse gas emissions.

Furthermore, kangaroos eat only about a third of the plant material that sheep consume and drink far less water, a consideration in Australia where droughts are a way of life. But kangaroos cannot be raised the same way as cattle or sheep. Because of their tremendous jumping ability, fences have to be about three metres high. Another problem is that the animals do not take well to captivity and develop *apture myopathy* in response to excess adrenaline release due to stress. This damages muscle and results in a bad taste. So the animals have to be hunted in the wild and, by law, are to be killed by a shot to the head. Kangatarians see this as a more humane practice than raising cattle, pigs and chickens on industrial farms and then subjecting them to distress in an abbatoir.

As one might expect, some animal rights groups are hopping mad about the kangaroo hunt, particularly about the "joey issue." These young kangaroos cannot survive, they claim, if their mothers are killed. Most hunters, however, shoot only the larger males, because their profits depend on the weight of the meat they bring in. In Australia, kangaroo consumption is increasing, but producers would like to make the leap across the ocean. Image is a problem— North Americans don't relate to eating the meat of an animal that looks like a giant bunny. That's one of the reasons a contest was organized in Australia to come up with a name that would be appealing when seen on a menu. They considered maroo, kangarly, kangasaurus, marsupan and MOM—for "meat of marsupial."

Why anyone would think that eating MOM would be an attractive proposition is hard to say. In the end, the winner was *australus,* although North Americans will probably look askew at australus burgers, if they do happen to appear. Eating kangaroo will probably be largely limited to kangatarians and Australian Aborigines, for whom the "roo" has supplied not only food, but clothing and recreation as well. Some tribes apparently use a stuffed kangaroo scrotum as a ball for a traditional game of *marngrook.*

What they do with the natural contents of the scrotum is not known. But I think if I were invited to a native feast in Australia and were offered "outback oysters," I would politely decline. That's because I have actually tasted prairie oysters in Alberta. Not a fond gustatory memory.

At a factory in India, workers douse animal dung with water that then seeps through a porous floor. The solution is collected and concentrated by boiling to eventually yield a chemical of great commercial importance. What is that chemical?

Potassium nitrate, commonly known as saltpetre. And no, it isn't being produced to curb anyone's libido. That's an old myth. Saltpetre doesn't have any anti-aphrodisiac properties, and it was never put into soldiers' food to dampen their sexual appetite. But it is an excellent fertilizer, supplying both potassium and nitrogen, key elements for plant growth. Along with sulphur and charcoal, it is also a key ingredient in old-fashioned gunpowder. By the Middle Ages, Europeans were already trying to obliterate each other with that flammable concoction invented by the Chinese sometime in the first millennium. Where did they get the necessary potassium nitrate?

The bat cave was one possibility. No, not Batman's cave, but real caves populated by real bats. Bat poop, or guano, contains a high concentration of potassium nitrate. But bats could not supply all the gunpowder that European countries needed in their efforts to beat each other into submission. Luckily, though—or perhaps unluckily—there was plenty of manure around, both animal and human. Dung heaps then became the prime source of saltpetre.

So, how do you go from dung to bang? Manure contains a variety of nitrogenous waste products such as protein residues and urea. If a manure heap is kept moist, preferably with urine, which adds its own supply of urea, the nitrogenous compounds are converted by bacterial action into ammonia. This then undergoes further bacterial oxidation to nitrate. If wood ashes, with their content of potassium carbonate, are stirred into the fermenting dung heap, the result is potassium nitrate. Being water-soluble, it can be leached out with water. That's how it was done in the

Middle Ages, and that's how it is still done in that factory in India. And where does the Indian potassium nitrate go? To feed the hungry. Not directly, of course. Saltpetre is hardly a nutritious delicacy. Unless, of course, you are a plant. Then saltpetre is a great fertilizer.

The Western world does not rely on manure for potassium nitrate. Modern production starts with ammonia, which in turn is made from hydrogen and nitrogen by the famous Haber process. Ammonia is then reacted with oxygen at high temperatures through the Ostwald process to yield nitric oxide, which reacts with more oxygen to form nitrogen dioxide. This in turn reacts with water and yet more oxygen to yield nitric acid. All that is needed to produce potassium nitrate is to neutralize this acid with potassium hydroxide. We've come a long way since Napoleon issued an edict that men were to urinate on the "nitre beds" that communities were expected to maintain. It seems the Little Emperor's conquest of Europe was aided by pee power.

What chemical naturally present in male saliva has been found to be reduced when men sniff female tears?

The concentration of testosterone, the male sex hormone that plays a role in sexual arousal and aggression, is reduced when men are exposed to odourless female tears. In a paper published in the prestigious journal *Science*, Noam Sobel and his group at the Weizmann Institute of Science in Israel report that levels of testosterone in male volunteers' saliva were found to be, on average, 13 per cent lower after they sniffed "emotional" tears secreted by women. The

research prompted a number of overly exuberant headlines such as
"A woman's tears are the biggest turnoff for men."

The biggest turnoff? Did the researchers compare the effect of
tears to that of bad breath, or BO, or gum chewing? No. So, what
did the research actually involve?

Sobel solicited volunteers who professed to be easily moved to
tears. No great surprise that, of the sixty respondents, only one
was male. And he didn't make it to the second round of selection,
which was based on the extent to which the tap was turned on
after watching a tear-jerking movie. Five women were judged to
be sufficiently proficient sobbers to supply tears needed for the
experiment. The first question was whether or not men could
distinguish between the odourless tears and a salt solution when
pads infused with either of these were pasted under their nose.
They could not. Next, the men were shown pictures of women with
emotionally ambiguous facial expressions and were asked about
what emotion they thought the women were feeling. Sniffing the
real tears had no effect on the men's interpretation of the women's
emotional status even when the men had been made to watch a sad
movie that supposedly increased their own emotional sensitivity.

That certainly was not headline material. But the researchers
also asked the men to rate the women's pictures in terms of
attractiveness. Of twenty-four male volunteer sniffers, seventeen
rated the women as being less attractive when they were exposed to
the emotional tears and reported reduced feelings of arousal.
Furthermore, they had their salivary testosterone level reduced by
an average of 13 per cent.

Although the study was small, involved tears from only a few
women, and the results were barely statistically significant, it was
attractive enough to generate catchy headlines like "Scent of a
Woman's Tears Lowers Men's Desire." These were certainly over
the top, but the study does present some features of interest. Could
it be that women's tears may actually manipulate male behaviour

through some sort of chemosignal expressed in tears? The tempting conclusion is that female tears shed in response to some sort of distress may cause a reduction in male aggression. Perhaps this is an evolutionary result of women protecting themselves against unwelcome sexual advances by men.

Another phase of the Weizmann Institute study supports the possibility of some sort of subconscious chemical signalling. When the male volunteers were subjected to functional magnetic resonance brain scans while sniffing female "emotional" tears, a reduced activity was noted in parts of the brain that are involved with sexual urges. Tears are chemically very complex, and specifically what compound leads to the testosterone reduction has not been determined, but obviously it has to be volatile, which lets out the various proteins, enzymes and minerals that have been identified in tears. The search is on to isolate the responsible chemical that appears to dampen sexual desire in men.

While headlines such as "A Woman's Tears—The Anti-Viagra?" are hardly justified by this study, it may not be a good idea for hopeful men to watch *Love Story* or *Terms of Endearment* in bed with their partner.

In 1907, Dr. Duncan MacDougall placed a series of six dying patients on a specially constructed balance. What was he trying to determine?

Dr. MacDougall, a reputable physician, believed in the existence of the human soul. There was certainly nothing novel about that belief; the idea that human life could not be totally explained by the physical workings of the body already had a long history. Indeed, most religions embrace the idea that after we are through using our body for our mundane existence, something of us survives and joins the Creator. This essence is commonly referred to as the soul. It is generally thought to be an immaterial entity—in other words, not something that can be seen, heard, touched or weighed.

But Dr. MacDougall had a different take on the soul. He contended that it was material. Whether it was located in the heart, as the ancient Egyptians believed; in the blood, as the Romans postulated; or, as according to some African tribes, in the liver, or maybe in the brain, he didn't know. But if it was a physical entity, it could be weighed. Since MacDougall believed that the soul leaves the body at the moment of death, he proposed a bizarre experiment. He would construct a balance capable of housing a literal deathbed and investigate any change in reading at the exact moment of death.

MacDougall managed to somehow find six patients at the very end stage of life—four suffering from tuberculosis, one from diabetes and one from unknown causes. One by one, he placed them on the deathbed scale and waited for their final breath. As death approached, there was a slow but steady loss of weight that he attributed to the evaporation of moisture. And then, at the very moment of demise, there was a sudden, larger loss of weight.

Could this have been due to the expulsion of the last breath? Apparently not. MacDougall confirmed this by lying on the bed himself and exhaling as much air as he could. In this case, the scale did not detect a change. The doctor therefore concluded that the weight loss at the moment of death was due to the soul leaving the body! He confirmed this in his own mind by repeating the experiment with fifteen dogs. There was no change in weight when they expired, meshing nicely with MacDougall's idea that, unlike humans, animals have no souls. How he managed to acquire fifteen dogs on the verge of death is not clear, but it is hard to imagine any process other than deliberately poisoning the animals. All for the sake of science, of course.

The results of this macabre "study" were published in the journal *American Medicine*, which, it seems, did not have a high regard for peer review. For there was plenty to criticize in this wacky experiment. How was the exact moment of death determined? How was the accuracy of the scale tested? How come there was no consistency in the results? In only one case was there a loss of weight at the supposed moment of death, with no further weight loss. In two of the cases, there was weight loss that increased with time. MacDougall's explanation was that in persons with "sluggish temperament," whatever that means, the soul may remain in the body for a full minute. When you have your mind made up, you can always rationalize. It may even be possible that different souls have different weights, or that some of us have no soul at all.

Certainly MacDougall's ethically and scientifically questionable experiments did not prove the existence of a soul. The accuracy of his equipment was suspect, the sample size too small and the results inconsistent. This nonsense cropped up again in the 2003 movie *21 Grams*, the title being a reference to the weight supposedly lost by MacDougall's most famous patient as he expired on his deathbed.

Too bad Dr. MacDougall is no longer around. I'd have a couple of questions for him. If the soul is indeed material, what is it made of? What is its chemical composition? Is it a single compound, or a mixture? Might different people have souls of different composition? If the soul is material, are soul transplants possible? And when the soul leaves the body, where does it go? The law of conservation of mass tells us that matter cannot be created or destroyed. Are souls recycled?

Why have baby bottles made of polycarbonate plastic been banned in Canada?

The concern was that bisphenol A (BPA) used to formulate polycarbonates could leach into the contents of the bottle. This controversial compound has been in the news a great deal because of its hormone-like properties. Back in 1993, Dr. Richard Sharpe of the University of Edinburgh's Human Reproductive Sciences Unit was the first scientist to suggest, in a landmark paper in the British medical journal *The Lancet*, that exposure during pregnancy to hormone-like chemicals could lead to developmental problems in the offspring. Since that time, his research has focused on male reproductive health and the role that environmental chemicals may play in the increasing incidence of testicular cancer, low sperm count and low testosterone levels.

Like most researchers in the field, Sharpe was open to the possibility that BPA, the chemical widely used in the production of polycarbonate plastics and epoxy resins, and which exhibits clear estrogen-like behaviour, could be a problem. After all, the chemical was showing up, albeit in trace amounts, in most everyone's urine!

But it was University of Missouri professor Fred vom Saal's finding of abnormal prostate growth and decreased sperm production in rats, at doses far lower than what was considered to be safe, that catapulted bisphenol A from obscurity onto the public stage.

Vom Saal's work, coupled with reports of female oysters acquiring male characteristics when exposed to the estrogenic detergent breakdown product nonylphenol and of penis deformities in Florida alligators linked to environmental contaminants, captured both the media's and scientists' attention. Grants became available, and research on BPA mushroomed to the extent that it may now well be the most studied chemical ever, with over 4,500 research papers devoted to it. One would think that would be enough to eliminate all confusion, but alas, such is not the case.

Reproducibility is one of the cornerstones of scientific research. But try as they might, other scientists were unable to reproduce the prostate problems in rats. Vom Saal argued that they were using the wrong breed of animals, or that the studies absolving BPA were funded by chemical companies and therefore could not be trusted. Actually, various breeds of animals were used, and furthermore, if subtle differences between closely related species have such a significant effect on bisphenol A metabolism, then how can we draw any inference about humans? And as far as studies are concerned, the relevant question is not the source of funding, but whether the data is robust enough to pass peer review.

The debate about BPA quickly took on an acrimonious nature, with accusations of "flawed studies," a usual euphemism for poor work, flying back and forth. But virtually all regulatory agencies came to the conclusion that there was no evidence of any harm to humans, although some, such as Health Canada, exercised the "precautionary principle" and banned polycarbonate baby bottles. That sat well with Dr. vom Saal, who commented: "The science is clear and the findings are not just scary, they are horrific. Why would you feed a baby out of a clear, hard plastic bottle—it's like giving a baby a birth control pill?" Well, the truth is that the science is not clear, and a comparison with the birth control pill is absurd. The estrogenic potential of the amount of bisphenol A that enters our body through our daily diet is about fifty thousand times less than that of one birth control pill.

This is not to say that there is no legitimate controversy concerning bisphenol A. Some researchers have detected blood levels they claim are comparable to those that cause problems in animals, but these studies have been criticized because they do not mesh with what is known about the amount of BPA ingested, how the chemical is metabolized, and the amount that shows up in the urine. Contamination of samples by traces of BPA in laboratory equipment has been offered as a likely explanation. To truly evaluate the effects of bisphenol A, it is imperative to determine blood levels accurately. And a recent study entitled "24 Hour Human Urine and Serum Profiles of Bisphenol A During High Dietary Exposure" did just that. Dr. Sharpe described the study as "majestic." The British do have a way with words.

The study was carried out jointly by a U.S. Department of Energy lab, the Centers for Disease Control and the FDA, not exactly fly-by-right operations. Twenty subjects consumed a diet of mostly processed foods that had been in contact with bisphenol A, as in can liners. According to calculations, this put them in the highest possible category of oral exposure to BPA. Blood samples were taken every hour over a twenty-four-hour period, a type of investigation that had never been conducted before.

Furthermore, the samples were sent for analysis of bisphenol A to two independent laboratories that were thoroughly experienced in preventing contamination. In a nutshell, the biologically active bisphenol A levels in the blood throughout the day were below the detection capabilities of the most sophisticated instruments available! This remarkably meticulous study casts doubt on previously detected blood levels by others.

"Beautifully designed and executed" was Professor Sharpe's comment about the study. His interpretation is that the majority of effects observed in animal studies can now be seen as probably not relevant to humans because they involved much higher BPA blood levels. Humans would have to be thousands of times more sensitive to BPA than rats to experience developmental issues, a very unlikely scenario.

Is this the final word on BPA? No. The final word in science is that there is rarely a final word. Undoubtedly, those who look on BPA as the chemical from hell are scrambling to poke holes in this new study. They may point out that it is not oral, but rather skin exposure to BPA that causes a rise in blood levels. Cash register ink is often made with BPA, and paper currency that is in contact with receipts may become contaminated. Indeed, almost all currency that has been investigated had traces of BPA, but the amounts were so small that researchers estimate exposure from such sources to be about ten times less than from house dust, which also contains trace levels of BPA. And of course BPA exposure from house dust would have been noted in the "majestic study."

A wagging finger may also be pointed at a recent study that found an association between higher blood levels of BPA and lower estrogen levels in the blood of women whose eggs were being harvested for in vitro fertilization. The implication is that this could have an adverse effect on the eggs. The health of the eggs was not investigated, and the study has not been reproduced, so any further conclusion is inappropriate.

I suspect that most readers have not heard of the "majestic study" I've been discussing. Isn't it amazing how any news that implicates bisphenol A as a potentially harmful substance makes headlines, but accounts of its safety get swept under the rug? Should a study show some sort of association between feeding some rodent an inordinate amount of bisphenol A and some hormonal change, it gets trumpeted in the press. Never mind that the study can't be reproduced by other researchers.

Or should there be a possible link between occupational exposure to bisphenol A and impaired sexual function, you can be sure that the headline writers will be at the ready with their clever quips. Never mind that such exposure has no relevance for the general population.

But when a major organization such as the German Society of Toxicology comes out with a peer-reviewed article in a prestigious journal, *Critical Reviews in Toxicology*, declaring that BPA poses no substantive risk, there is nary a mention. Except perhaps some bluster in the Huffington Post about all the members of the German panel being allied to the plastics industry, which is blatantly untrue.

Neither do I recall seeing headlines when the U.S. Food and Drug Administration reviewed the chemical in 2010 and concluded that it was safe at current low levels of exposure, or when the FDA ruled in 2012 that no new regulations were needed to further limit bisphenol A in foods. Neither was there great press coverage when the European Food Safety Association examined over eight hundred studies and concluded that BPA was not harmful. Indeed, after some fifty years of use, and thousands of studies, there is no evidence that bisphenol A harms human health.

Of course, it is true that the absence of evidence of harm is not evidence of the absence of harm. Given the fact that this chemical does have hormone-like properties, and in test tube experiments and animal trials can have some endocrine-disruptive effects, the possibility of subtle effects on humans cannot be ruled out. This would seem to be especially the case during early development. If BPA is to cause any harm, it is most likely to do so if exposure is in utero or during infancy.

It is almost dogma that scientific papers wrap up with the statement that "more studies are needed." But, as blasphemous as that may sound, perhaps with BPA, enough is enough. After years and years of research and buckets of money spent, we have learned that if there is any danger it is vanishingly small, otherwise it would have been noted. Indeed, that is the conclusion that Dr. Richard Sharpe has now come to. Here is what he says: "Fundamental, repetitive work on bisphenol A has sucked in hundreds of millions of dollars and it looks increasingly like an investment with nil return. All it has done is to show that there is a huge price to pay when initial studies are adhered to as being correct when the second phase of scientific peer review, namely, the inability of other laboratories to repeat the initial studies says otherwise."

He references a study in the journal *Toxicological Sciences* that showed complete absence of effect of a range of bisphenol A exposures on reproductive development, function and behaviour in female rats. The results of this study are robust and unequivocal and counter previous, more circumspect studies that spawned the BPA fear. The press, however, didn't pick up this story.

It may never be possible to tease out the effects, if there are any, of BPA on human health, because the fact is that our bodies are awash in both natural and synthetic compounds that are potential endocrine disruptors. Soy, flax, wine, nuts and beer all contain such, and then there are various plasticizers, pesticides and disinfectants. One can argue that finding substitutes for BPA in food applications is an appropriate application of the precautionary principle. Maybe. But the evidence to support it is pretty flimsy.

Professor Sharpe, who you'll remember was once on the anti-BPA bandwagon, now believes that the supposed risk of BPA was arrived at based on faulty studies. He opines: "As scientists, we all like our ideas and hypotheses to be proved correct; yet there is equal merit in being proved wrong." The evidence now suggests that the notion that bisphenol A is a chemical from hell is wrong.

The time has perhaps come to put estrogenic concerns about this chemical on the back burner. Of course, that back burner needs to be monitored for flare-ups.

Why are some accountants and bookkeepers taking to wearing latex gloves when preparing their clients' income tax submissions?

Here too we are dealing with bisphenol A. The thermal printing paper that is commonly used for cash register receipts contains bisphenol A, the chemical that has generated a great deal of controversy because of concerns that it may have an effect on our health even at very low concentrations. That controversy has generated a great deal of ink in newspapers and magazines, and that ink, too, is often formulated with BPA. Is it possible that just handling cash receipts or newspapers or magazines can transfer BPA to the skin, and that from there it can find its way into the bloodstream?

Science can quite readily answer those questions, and it has. Indeed, BPA can be absorbed by the skin through handling paper that contains BPA, and it can then pass into the bloodstream. Of course, whether the amount that ends up in the bloodstream is consequential or not is a totally different question, and one that is much more difficult to answer.

How do we know that BPA is absorbed by the skin? Sometimes, pigs' ears come in handy for more than serving as tasty chews for dogs. Pig ear skin is a widely used model for human skin. Dr. Daniel Zalko at the French National Institute for Agricultural Research treated pig skin with radioactively labelled bisphenol

A and followed the movement of the chemical through the skin by monitoring radioactivity. About 65 per cent of the BPA diffused through the pig skin. This prompted experiments with human skin, which showed that about 46 per cent of the chemical made it through. That still left the question of whether or not this can lead to an increase in blood levels, since the possibility remained that passage through the skin allowed for BPA to be metabolized. It turns out, however, that the compound does make it through the skin intact. Dr. Joe Braun at Harvard looked at the concentration of BPA in the urine of 389 pregnant women who worked as cashiers, teachers or had industrial jobs. The cashiers who handled a lot of thermal paper had the highest levels of BPA, suggesting that BPA is absorbed through the skin and does end up in the circulatory system, eventually to be eliminated in the urine.

Researchers in Switzerland added some corroborative evidence when they had volunteers hold thermal printing paper for five seconds and then analyzed the skin for BPA. About one microgram was transferred, with the amount depending on whether the fingers were dry, wet or greasy. Wetness or greasiness was associated with ten times greater transfer of BPA to the skin. In an attempt to monitor the fate of the BPA, the researchers investigated whether the BPA absorbed into the skin was extractable with alcohol after two hours.

It turned out that after holding the thermal paper, the BPA was extractable from the skin after two hours, but when applied in an alcohol solution it was no longer extractable. This suggests that BPA can pass into the bloodstream depending on skin conditions. Based on the amount that was not extractable, the researchers concluded that exposure to someone touching thermal paper for ten hours a day could lead to the absorption of 70 micrograms of BPA per day, which is some forty-two times less than the present tolerable daily intake. However, if there is cream on the hands, the margin of safety could be less.

There is also the question of where the thermal paper eventually ends up. Much of it heads for recycling, meaning that facial tissues and toilet paper could also lead to BPA exposure. It should also be noted that in thermal papers and in recycled papers, BPA is a free compound, as opposed to its presence in polycarbonate plastics or can liners where it is incorporated into polymers. Absorption from paper, therefore, is easier. The bottom line is that it is not a bad idea, especially for young women working as accountants or cashiers who may become pregnant, to wear latex gloves when handling receipts. Then all they have to worry about is latex allergy.

You're going to need some iron supplement tablets and some black tea. What are you trying to make?

We wouldn't have much recorded history if it weren't for ink. After all, carving into stone tablets is not a very efficient recording system. The Chinese used plant dyes and ground charcoal as early as the eighteenth century BC, and in India, "India ink" was made from burned bones, tar and pitch. The Romans and Greeks concocted mixtures of soot and gelatin, but the classic ink of antiquity, likely first produced by the Romans, was "iron-gall ink." Naturally occurring iron sulphate was mixed with a solution made by steeping oak galls in hot water. Galls are growths on the bark of the oak tree caused by a type of wasp laying its eggs in the bark. These growths are rich in tannic acid, the critical component for making ink. It is the tannic acid that reacts with iron to make an indelible dark complex.

While oak-gall ink is permanent, which is desirable, it can fade into a brownish colour. But more importantly, the ink is corrosive to parchment and paper. Many manuscripts written with iron-gall ink are now threatened with corrosion. These include some of the writings of Leonardo da Vinci and Johann Sebastian Bach. Indeed, Bach's works are already in a state of advanced decay.

A version of gall ink can be made with tea, which contains a high concentration of tannins. A strong tea is needed, which can be made by steeping a tea bag in a couple of tablespoons of boiling water. Ferrous sulphate is readily available as a dietary iron supplement, and the pill can be easily crushed with a pestle in a mortar. As soon as the ground powder is added to the tea, it will turn to "black ink."

When you brew tea for drinking, you are also making some ink, since water contains some dissolved iron. The darkness of the tea depends on how much tea you use and the amount of iron in the water. If you use water that has been passed through a filter that removes some minerals, such as a Brita filter, the tea will be a lighter colour due to a lower concentration of iron. But you can also lighten your tea by adding lemon juice. Citric acid breaks down the iron-tannin complex. Sometimes, water filter promoters demonstrate that filtered water makes for a lighter-coloured tea, implying that the water is purer than tap water.

While it is true that the water may be purer, the lightness of the tea does not demonstrate a health advantage. It only shows that the filter has removed some iron from the water, which is not necessarily desirable, since iron is a nutrient we require.

What is silver doing if it is exerting an "oligodynamic" effect?

Discovered in 1893 by the Swiss botanist Karl Wilhelm von Nägeli, the oligodynamic effect refers to the toxic effect of metal ions on living organisms such as bacteria, algae and fungi. Copper, lead, zinc, gold and mercury can also furnish ions that will produce an antimicrobial effect, but the toxicity of lead and mercury preclude their use in this fashion. That wasn't always the case. Mercury at one time was commonly used to treat syphilis, giving rise to the expression "A night with Venus, a lifetime with mercury." However, because of mercury's toxicity, that lifetime was often quite short. Copper and silver, on the other hand, have low toxicity by comparison and can be effectively used to battle bacteria.

As early as the ancient dynasties of Egypt, silver coins were placed in the drinking vessels of the nobility to protect them from harm. Of course, this was not the result of any scientific investigation; the practice probably originated from some superstitious belief about the magical properties of precious metals. Over the years it became apparent that the silver coins really did have an effect: they kept water from becoming slimy.

Storage of water in silver vessels was an obvious extension of this observation, affording the well-to-do some protection from waterborne diseases that were common before the introduction of chlorination. As recently as the twentieth century, when Maharaja Sawai Madho Singh II journeyed to Britain, he carried holy water from the Ganges River in two sterling silver vessels acclaimed to be the world's largest silver containers.

You do not have to be as rich as a maharaja to experience the oligodynamic effect of silver. The metal can be incorporated into urinary catheters and endotracheal breathing tubes to reduce infections, and fabrics can be formulated with small amounts of silver to control the bacteria responsible for churning out odorous compounds when they feast on sweat. The antimicrobial effect is actually due to silver ions—in other words, silver atoms that have lost an electron. These ions are produced whenever silver atoms at the surface of the metal react with either oxygen or hydrogen sulphide, the "rotten egg" compound that is always present in the air and water in trace amounts. Indeed the tarnish on silver is silver sulphide, the product of the reaction between silver and hydrogen sulphide.

Silver ions inactivate enzymes that are essential for bacterial life. That's why bacteria are killed when contaminated water is stored in a silver container. But in this case, the extent of disinfection is unreliable because the concentration of silver ions in the water cannot be controlled. The purity of the silver, the size of the container, whether the water is shaken—all are important determinants of the concentration of silver ions. However, techniques have been worked out to produce just the right concentration of ions by immersing a pair of silver electrodes connected to a direct current into water that needs to be purified. This was the method used to produce drinking water aboard the Apollo space flights and is used in hospital plumbing systems to deactivate legionella bacteria. Copper is often alloyed with silver in the electrodes to take advantage of its oligodynamic effect as well. Swimming pool disinfection systems using copper-silver ionization that allow for reduced use of chlorine are also available.

If silver ions produced on the surface of the metal are the active disinfecting agent, it stands to reason that the surface area of the silver would play an important role in the effectiveness of the treatment. And it does. Particles of silver that are less than a billionth of a metre in size, commonly referred to as *nanoparticles*, have been shown to be especially effective at killing bacteria. A recent study by Dr. Derek Gray at McGill University showed that passing contaminated water through absorbent blotting paper treated with silver nanoparticles resulted in inactivation of bacteria. This has the earmarks of a landmark discovery, because the consumption of contaminated water in the developing world is a major health crisis. Nanosilver-impregnated paper is easy to produce and easy to use.

Unfortunately, quackery often rides the coattails of science. And so it is in this case. Numerous websites promote the use of "colloidal silver" as a cure for cancer, diabetes, HIV infection and herpes. Colloids refer to a system in which finely divided particles are dispersed within a continuous medium without settling out. In the case of colloidal silver, these particles can be elemental silver or particles of silver compounds.

Indeed, they may well have an antibacterial effect in a Petri dish, but that is a long way from having an antibacterial effect when taken internally. No scientific evidence supports the benefit of ingesting any form of colloidal silver. Making health claims on its behalf is illegal, but colloidal silver can be sold as a dietary supplement. That is a curiosity, because humans have no dietary requirement for silver and there is no such thing as silver deficiency.

But there certainly is such a thing as silver excess. The metal can deposit in the skin as well as in internal organs, with the result being a condition known as *argyria*. Its hallmark is grey-blue skin, a condition that is irreversible. One of the most famous cases was that of the Blue Man who was a featured attraction in Barnum and Bailey's circus in the early years of the twentieth century. He had apparently tried to cure his syphilis by ingesting silver nitrate, but succeeded only in making himself blue.

More recently, Stan Jones, an American Libertarian who twice ran unsuccessfully for the U.S. Senate as well as for governor of Montana, did succeed in becoming blue. Back in 2000, he was worried that computers would stop functioning and that this would somehow make antibiotics unavailable. He decided to take preventative action and started to take a colloidal silver preparation he made himself by passing an electric current through a solution equipped with silver electrodes. Unfortunately, he didn't know what he was doing and used too high a voltage, and his solution contained a great deal of silver. He turned blue. But he is not singing the blues, maintaining that he is healthy. He still dopes himself with colloidal silver. Pretty dopey, actually.

What is the cause of both gilder's palsy and "hatter's shakes"?

Physicians today are unlikely to encounter gilder's palsy. Nor are they likely to diagnose a patient with hatter's shakes. But prior to the twentieth century, these ailments had to be considered when a patient presented with tremors, irritability, increased salivation and fatigue. The culprit was mercury. In the case of the hatters, it was mercury nitrate used to produce felt. Beaver and rabbit fur, the traditional materials for making felt, can be matted more easily when the pelts are first treated with mercury nitrate, a chemical that opens up the pine cone–like layers, known as *imbrications,* on the surface of individual hairs. When these are opened up, adjacent hairs can interlock more readily. Hatters invariably got the mercury nitrate on their hands, and since their hygiene was probably less than exemplary, ended up ingesting some of the toxin.

Mercury's toxicity is a consequence of its ready binding to sulphur, an element that is a crucial component of many enzymes. Some of these enzymes are critical to the workings of the central nervous system, and their failure to function properly when bound to mercury causes the shakes and mental disturbances that are characteristic of mercury poisoning.

Gilders, whose profession was based on coating metal objects with gold, exhibited symptoms similar to those of hatters. Their problems, however, came not from exposure to compounds of mercury, but from exposure to metallic mercury, the silvery liquid found in thermometers. The Romans called the metal *hydrargyrum*, meaning "liquid silver." That also explains why we use the symbol Hg for the element. Unlike mercury nitrate, liquid mercury is somewhat volatile and can therefore be inhaled and absorbed into the bloodstream from the lungs.

Metallic mercury does not occur in nature, but it can be produced by heating cinnabar, a naturally occurring form of mercury sulphide (HgS). The metal has long fascinated people, especially the alchemists who thought it was the key to the transmutation of base metals into gold.

Of course it was not that, but there is a gold connection. Gold readily forms an alloy with mercury, a phenomenon that is apparent to anyone who has handled mercury while wearing a gold ring. While playing with mercury is a bad idea, the historical alloying with mercury to form gold amalgam has been an important method for isolating gold from ores. The traditional process involves crushing the gold ore, mixing it with mercury and separating the amalgam that forms. This is then heated to drive off the volatile mercury, leaving pure gold behind. But it can leave something else behind as well: the misery of mercury poisoning. And many a button gilder could have testified to that.

Military uniforms commonly feature golden buttons. Until about the middle 1800s, these were made by dipping metal buttons into gold amalgam and then heating to evaporate the mercury. The layer of gold left behind was very thin; just one gram of gold was enough to gild about five hundred buttons. The results for the buttons were pretty, but for humans, not so much. On occasion, even construction workers had to deal with gilder's palsy. About a hundred kilos of gold were mixed with mercury for application to the copper sheets that were used to create the golden dome that adorns the cathedral of St. Isaac in St. Petersburg. The dome, unfortunately, is also a symbol of mercury poisoning. Some sixty workers died as a result of mercury inhalation. However, the chemical ingenuity of two Italians would eventually put gilder's palsy on the back burner.

In 1800, Alessandro Volta's discovery of an electric current flowing between two dissimilar metals separated by moistened cardboard established the chemical principles that would lead to the first battery. Just five years later, his friend Luigi Brugnatelli reported in the *Belgian Journal of Physics and Chemistry* how he had put the "voltaic pile" to use: "I have lately gilt in a complete manner two large silver medals, by bringing them into communication by means of a steel wire, with a negative pole of a voltaic pile, and keeping them one after the other immersed in ammoniuret of gold newly made and well saturated." Brugnatelli had discovered electroplating, a process that would be commercialized by Henry and George Elkington of Birmingham, England, in 1840.

The elimination of gilder's palsy thanks to electroplating did not mean that we were rid of poisoning from inhaled mercury. Consider the case of the sixty-eight-year-old man and his eighty-eight-year-old mother-in-law who were admitted to hospital in Michigan with nausea, diarrhea and vomiting. The next day, the man's son and daughter-in-law were brought in with the same symptoms. Because the esophagus and lungs of all four were inflamed, doctors suspected chemical exposure. It turned out that the son worked for a company that manufactured dental amalgam, which is an alloy of mercury and metals, mostly silver.

With the price of silver rising, he had the bright idea of stealing some of the material and extracting the silver by evaporating off the mercury. Thinking he had a formula for riches, the prospective alchemist set up a crude lab in the basement, equipped with a furnace for melting metal. He had a formula, all right, but it was one for disaster. Despite the use of dimercaprol, a drug that can bind to mercury through its sulphur atoms, all four mercury-poisoning victims died. Their house was demolished and the debris treated as hazardous waste.

You don't have to be a greedy crook to suffer from mercury inhalation. You can be an inquisitive youngster. Like the boy who dissected a household thermostat and spilled the mercury on the carpet. He vacuumed it up, but never changed the bag. With each subsequent use of the vacuum cleaner, some of the mercury vapourized. Months later, the boy was hospitalized with weakness, weight loss, anorexia, lethargy and insomnia. Luckily in this case, dimercaprol treatment was effective. Too bad Alice didn't have any to offer the Mad Hatter.

What would be the product of reacting acetone with hydrogen peroxide and sulphuric acid?

Terrorists are familiar with this combo of reagents, which are used to make the explosive triacetone triperoxide (TATP). This is the explosive that originally prompted the airport screeners' question, "Do you have any liquids, gels or powdered fruit drinks?" Except for the powdered fruit drinks, such questions have become routine. But on July 10, 2006, I had no idea why I was being asked this bizarre question. Why would the agent be concerned about my toiletry and dietary habits? I couldn't make heads or tails of it. The only connection with flight that sprang to mind was with Tang, the orange-flavoured crystals that John Glenn took along on his orbital mission in 1962. But we were travelling from London to Budapest, a trip that was presumably not going to take us through outer space.

Having just spent a harrowing day at Heathrow, queueing for bathrooms and fighting for food after the cancellation of all flights, I gave a curt no for liquids or gels, and muttered something about powdered fruit drinks not being my cup of tea. At the time, all I knew was that the massive disruption was caused by some sort of terrorist threat. Only when we got to Budapest did I hear that the threat had something to do with a "liquid bomb." And, amazingly, with Tang.

My chemical curiosity was of course aroused, but the matter took on a personal touch when I learned that Air Canada's London–Montreal flight, the one we were going to take back home a week later, was one of the ones targeted. Which specific day the terrorists had chosen to try to blast seven planes out of the sky was not clear, but the attacks were apparently imminent, judging by the fact that the terrorists who had been under extensive surveillance for a month were suddenly arrested on the eve of July 6.

The whole caper began when British security secretly opened the baggage of Ahmed Ali Khan as he returned from Pakistan. Khan had raised some red flags because of his hard-line anti-Britain political stance, and when his suitcase was found to contain a large number of batteries and a supply of Tang, officials decided to mount a surveillance operation. It seemed unlikely that Khan was into battery-powered gizmos or that he was bent on cleaning his automatic dishwasher (Tang, because of its citric acid content, is great for that), or that he had developed such a fondness for orange-flavoured coloured water that he had to take a supply of Tang on foreign trips.

After one of Khan's associates was seen disposing of empty bottles of hydrogen peroxide, a video camera was secretly planted and caught the men constructing some sort of device out of beverage bottles. When Khan was seen checking out flight schedules at an Internet café, the decision was made to arrest him and his bunch. Details of this operation were not released, but reporters somehow got wind of hydrogen peroxide and Tang being involved.

And then the speculation started. Newspaper accounts proposed that the terrorists were actually going to make a bomb from chemicals smuggled through security disguised as beverages by colouring with Tang. Acetone, hydrogen peroxide and sulphuric acid would be used to make the powerful explosive TATP. The necessary materials would not be hard to acquire. Acetone is readily available as nail polish remover, sulphuric acid is the acid in car batteries, and the concentrated hydrogen peroxide needed can be made by boiling off the water from the 3 per cent peroxide sold in pharmacies.

Indeed, TATP can be made quite easily by a competent chemist, and it has been used in many a suicide bombing. But synthesis requires careful temperature control, mixing, filtering and drying, hardly the operations that could be carried out in an airport or airplane toilet.

As more information came to light during the trial of the terrorists, other possible scenarios emerged. Apparently, one of the videos taken at the "bomb factory," as the house where the gang met was dubbed, had shown Khan drilling a hole in the bottom of a bottle with syringes and battery casings nearby. The exact details of what the men were doing and the various chemicals found after the arrest were described to the jury but were not made public.

Speculation was that the fruit crystals dissolved in concentrated hydrogen peroxide were to be introduced by means of a syringe through the bottom of a bottle that had been emptied by the same means. Hydrogen peroxide is an excellent source of oxygen, and the sugar in the powdered beverage can serve as a fuel, setting the stage for an explosion. All that is needed is a detonator, which can be made by filling a hollowed-out battery with hexamethylene triperoxide diamine (HMTD). Sounds like a complicated task, but HMTD can be made from hydrogen peroxide, ammonia and formaldehyde. Khan and his fellow terrorists could have done this.

Supposedly, the idea was to fill a bottle with the explosive mixture, seal the hole at the bottom with Krazy Glue, and take it aboard the plane as a beverage. At the appropriate time, the cap would be removed and the detonator-filled battery shell dropped into the bottle. The explosion would then be triggered with a jolt of electricity from a camera. The jury was, in fact, shown a video of an explosives expert carrying an orange-coloured drink into the mock-up of an airplane fuselage and causing a devastating explosion. What exactly was in the bottle we do not know, because the technical details were only for the eyes and ears of the jury.

The accused admitted to making explosives, but claimed they were just going to set them off in a public place to make a political statement without causing harm. Defence lawyers pointed out that, in fact, no flight reservations had been made. But they couldn't explain away the suicide videos found after the arrests and threats of jihad uttered. The jury was convinced of the men's guilt, and the major players were sentenced to life in prison.

So, could they have pulled it off? I've looked into the chemistry in much greater detail than I described here, for obvious reasons. Let's just say that the next time you are asked if you have liquids or gels, be glad they're asking.

Fit a sock that has been soaked in a solution of sugar and potassium chlorate around a bottle filled with gasoline and concentrated sulphuric acid. What have you made?

A Molotov cocktail. The Molotov cocktail is a rather primitive but effective incendiary device disdainfully named after Vyacheslav Molotov, Soviet prime minister under Stalin. Prior to the Second World War, under Molotov's direction, the Soviets attacked Finland in order to secure port facilities. The Finns were ill equipped to fight against the Soviet tanks and had to resort to homemade weapons. A simple one was made by filling a bottle with gasoline and fitting it with a rag also soaked in gasoline. Lighting the rag and then smashing the bottle against a tank caused a fireball and a raging fire. Some of the gasoline penetrated through openings into the tank, setting the insides on fire. But throwing a Molotov cocktail was a risky business. Sometimes, the bottle would ignite prematurely in the hand. The Finns did manage to develop a more sophisticated version to avert this problem. Instead of lighting the rag, they secured a couple of all-weather matches to the outside of the bottle. When the bottle hit the target, friction ignited the match, which then ignited the gasoline. If all went well.

Polish resistance fighters "improved" upon this idea during the war by making use of a chemical reaction. A mixture of sugar and potassium chlorate can be readily ignited by the addition of sulphuric acid. Sugar serves as the fuel, and potassium chlorate is an oxidizing agent, meaning it can furnish the oxygen required for combustion. The heat needed to get the combustion under way is supplied by the reaction between sugar and sulphuric acid. Sugar is a carbohydrate, so named because its chemical formula suggests it is composed of carbon and water. Sulphuric acid is a powerful dehydrating agent, meaning it can strip water out of carbohydrates. This reaction produces a great deal of heat, enough to light the potassium chlorate–sugar combo. The idea then is to soak a sock or rag in sugar and potassium chlorate and tie this around a bottle full of gasoline with some added sulphuric acid. When the bottle breaks, the acid combines with the sugar and the chlorate, a fire is started and the gasoline ignited. Later versions incorporated some motor oil or chopped-up Styrofoam to make the mixture sticky and even more deadly. Quite ingenious, really.

✲

It contains either ammonium phosphate, sodium bicarbonate or potassium bicarbonate. You don't eat it. What do you do with it?

Put out fires. Ammonium phosphate, sodium bicarbonate or potassium bicarbonate are the ingredients found in dry chemical fire extinguishers. Three conditions have to be met for a fire to be sustained. Fuel has to be available, oxygen has to be present and a high temperature has to be maintained. To extinguish a fire, this "fire triangle" has to be broken. An extinguishing agent does this either by lowering the temperature, by absorbing heat or by blanketing the fire with some substance that prevents contact with oxygen. Water cools a fire very efficiently, but it cannot be used on electrical fires since it conducts electricity. Nor can it be used on flammable liquids, which can float on top of the water and actually spread the flames. Basically, water is only effective on what is referred to as a Class A fire, meaning that the fuel is wood, paper or fabric.

But when flammable liquids are present, as in a Class B fire, or when live electrical circuits are involved, as in a Class C fire, dry chemical extinguishers are needed. Powdered sodium bicarbonate is the simplest dry chemical used. When propelled out of the extinguisher by pressurized carbon dioxide, it quickly covers the fire and shuts off the oxygen supply. An additional feature is that heat causes sodium bicarbonate to break down, releasing carbon dioxide gas, which, being heavier than air, helps to exclude oxygen.

Sodium bicarbonate gets blown around too easily when paper or wood fires are involved, and is therefore not very effective as an extinguishing agent. But ammonium phosphate is. It melts and covers the fire with a thin layer of viscous liquid that cuts off the oxygen supply. A dry chemical extinguisher that is rated as effective for Class A, B and C fires usually contains ammonium phosphate. This technology has come a long way since 1723, when Ambrose Godfrey invented the first fire extinguisher. That device contained a canister of gunpowder inside a metal container filled with water. Lighting a fuse caused the gunpowder to explode and spray water everywhere. It likely worked on fires, but one wonders how many firefighters were also extinguished in the process.

What is the link between paper cups, dental floss, early sound recordings and a Brazilian plant?

Carnauba wax. The leaves of the carnauba palm, native to Brazil, are covered with a wax that has a variety of uses. It is also helpful that separating the wax from the leaves does not require sophisticated technology. The leaves are cut into small pieces, allowed to dry and beaten. The wax flakes off and can be compacted into blocks for shipping. Like any natural wax, it is a mixture of compounds, but the main components are organic molecules called esters. There are also alcohols and fatty acids, along with some cinnamic acid.

Waxes are resistant to water, and can be buffed to a shine or act as lubricants. Carnauba wax is commonly used to polish cars, floors, furniture and shoes. Because it is hypoallergenic and doesn't allow water to pass through, carnauba is used in cosmetics such as lipsticks and cosmetic creams. The wax's resistance to water also allows it to be an effective coating for paper cups. It also protects and adds shine to Tic Tacs, jelly beans and apples. Carnauba's lubricating properties are used in the production of waxed dental floss. Since carnauba is the hardest of all natural waxes, it played a role in early sound recordings.

The very first sound recording by Edison in 1877 was on a cylinder made of brass and coated with tin foil, but this was not suitable for mass production. A stylus, in response to changes in air pressure, which is what sound is, engraved grooves in the soft metal. The grooves featured valleys and peaks that could reproduce the original sound when traced with a needle connected to a diaphragm. The diaphragm vibrated in response to the up-and-down movement of the needle, and when it was mechanically connected to a speaker, sound filled the room. In 1881, Chichester Bell and Charles Sumner Tainter developed a cylinder made of carnauba wax over a cardboard base, which made mass production possible.

Edison almost immediately improved upon this discovery and introduced the Improved Phonograph, which played wax cylinders. Eventually, he introduced cylinders made of plastic—celluloid in 1901, "Condensite" in 1908, and nitrocellulose in 1912. By this time, cylinders were being replaced by gramophone records, invented by Emile Berliner in 1885. His early gramophone discs, made of beeswax, featured a groove etched by sideways vibrations of a stylus. In 1889, beeswax was replaced by Vulcanite—a hard rubber—and then by shellac. Eventually, polyvinyl chloride, the vinyl of the record industry, made the mass production of long-playing records possible. The recording industry has come a long way from carnauba wax, but the wax is still around, adding lustre to our lives in various ways.

Flexible flip-flops and hard water pipes can both be made from polyvinyl chloride. How can the same plastic be used to make both hard and flexible items?

Flexibility is achieved by the addition of substances known as plasticizers, of which the most common ones by far are the *phthalates*.

These compounds may make plastics flexible, but they also generate some inflexible opinions about their safety. Some points, though, are beyond argument. There is no contention about the size of the industry. It is huge! Millions of pounds of various phthalates every year are produced for incorporation into a plethora of consumer items ranging from floor tiles and shower curtains to dashboards and toys. Since phthalates are not chemically bound to the plastics they soften, there is no argument about the fact that they can leach out. And since vinyl products are ubiquitous, it isn't surprising that phthalates show up in house dust, in food and in our blood.

Of course, the pertinent question is whether this matters. Presence of a chemical does not equate with risk, but neither can that possibility be dismissed just because the amounts detected are very small. In some cases, especially when it comes to substances that may have hormone-like behaviour, even tiny amounts may exert a physiological effect, which may be subtle and difficult to identify conclusively.

There is no doubt that in some species of rodents, some phthalates can mimic male hormones and cause abnormalities in the animals' genitalia and reproductive systems. No such effect has been shown in humans, save for one contentious study that claimed an association between higher phthalate levels in a mother's blood and a shortened distance between the anus and genitals of their male offspring. Even in this case, the researchers found no health consequences other than this curious observation. Indeed, a recent study of young adults who, while hospitalized twenty years earlier, had been exposed to high doses of phthalates leaching from flexible plastic tubing showed no adverse effect. On the other hand, some studies have found an increased risk of asthma associated with phthalates in house dust.

The vinyl and phthalate industries claim that scientific research supports the safety of phthalates and that, in some fifty years of use, there has not been a single known case of anyone having been adversely affected. When an investigation reveals, as was recently the case, that vinyl flip-flops can harbour as much as 23 per cent phthalates by weight, the industry quickly generated data to ridicule the notion that anyone could be harmed by the presence of these chemicals. Even if a person wore these flip flops twenty-four hours a day, they calculate, the amount of phthalates absorbed through the skin would be way less than the maximum amount that has been shown to cause no effect in test animals.

True, but hardly the point. Nobody claims that wearing plasticized flip-flops presents a danger to the wearer, unless he were to eat the shoes, and probably not even then. The concern is about where the phthalates in the shoe end up as the shoe wears away or is discarded. Are those chemicals biodegraded, or can they come back to haunt us in subtle ways by showing up in the food we eat or in the air we breathe? Industry spokespeople and a number of scientists dismiss this possibility, while activists are prone to exaggerating and twisting the animal data to come to absurd conclusions about "toxic" vinyl erasers and "slow death" by vinyl bathtub duckies.

Where does this leave the rational scientific thinker? Talk of banning vinyl or phthalates is sheer nonsense. These chemicals are far too useful, and the products in which they are found contribute immensely to the quality of our life. But not all phthalates are created equal, and not all their uses are equally necessary. The phthalates that have raised the most concern can, and are, being eliminated from toys. Their inclusion for the sole purpose of retarding the evaporation of fragrance in products to be used on babies' skin is unnecessary. But in the overall context of all the chemicals we are exposed to, both from natural and synthetic sources, in the game of life, phthalates are minor players.

✫

A tragic spill of red mud, a byproduct of aluminum production, resulted in several deaths and widespread environmental contamination in Hungary. What makes the mud red?

The red is due to various iron compounds, mostly oxides of iron, familiar to us as *rust*. Aluminum is produced from bauxite, a mix of minerals composed mostly of aluminum oxide and iron oxide, with small amounts of titanium dioxide and traces of other metals. There is no scarcity of bauxite; aluminum, in fact, is the most abundant metal in the earth's crust. But its journey from ore to usable metal is a long, arduous one. First, aluminum oxide has to be separated from its contaminants in bauxite, and then it has to be stripped of its oxygen content. These are not simple tasks.

Bauxite is mined in open pits and begins its trip towards pure aluminum by being subjected to the Bayer process. It was in 1888 that Karl Josef Bayer developed the process that has become the cornerstone of aluminum production around the world. At its heart is the reaction of ground, slurried bauxite with sodium hydroxide to form soluble sodium aluminate. The insoluble impurities, namely iron oxides, can now be separated by filtration. This is the "red mud" that has no commercial use and is stored in giant holding facilities. The sodium aluminate solution is then concentrated, resulting in the crystallization of aluminum hydroxide, which is calcined in rotary kilns to yield pure aluminum oxide. Dissolving this oxide in molten cryolite (sodium hexafluoroaluminate) and passing an electric current through the solution yields metallic aluminum, which is then cast into ingots. It is these ingots that are melted down and used to make a myriad of aluminum products ranging from engine blocks and window casings to electrical wiring and food cans.

Since roughly four tons of bauxite are needed for every ton of aluminum produced, there is obviously a lot of red mud to contend with. While it contains iron and titanium, both useful metals, there is no economic way to extract these. Residues of sodium hydroxide—*lye*, as it is commonly known—make the mud caustic and difficult to deal with. It ends up being stored in reservoirs until enough water evaporates, leaving a residue that can be buried.

The accident in Hungary occurred when one of these reservoirs failed and released its content of toxic sludge, sweeping across the countryside before spewing its highly alkaline contents into the Danube River. At least five people were killed and more than a hundred treated for burns and other injuries. Fish and plant life in the immediate vicinity also suffered a devastating blow, and thousands of acres of agricultural land will probably have to be dug up to get rid of the alkaline sludge.

Further problems may arise from arsenic, chromium and mercury, which are also present as contaminants in red mud. Not only can these contaminate water, they can be present in the airborne dust produced when the red mud dries. What caused the walls of the reservoir to give way is not known.

Where would you find polyfluoroalkyl phosphate esters?

Of course, you could find them in a chemical laboratory. But you could also find them in that laboratory we know as the human body. And where do they come from? A gazillion types of food packaging materials, ranging from your hamburger wrapper and microwave popcorn bag to your pizza box, are treated with these chemicals to make them greaseproof. Right, we don't eat the packaging, although in some cases that might be healthier than eating the food they contain. But polyfluoroalkyl phosphate esters, or PAPs, do migrate from the packaging into the food, and hence into our bloodstream. We know this because PAPs can be detected in human blood serum. Of course, just because they are detected doesn't mean that they are causing any harm. The important question that arises concerns the fate of these chemicals. And now, thanks to the elegant work of Dr. Scott Mabury at the University of Toronto, we have evidence that they are biotransformed, at least to some extent, to perfluorocarboxylic acids, such as perfluorooctanoic acid (PFOA). This is where we run into a situation that raises some eyebrows.

PFOA has been a concern since the 1970s, when it was first detected in human blood. The source was assumed to be the large-scale manufacture of fluoropolymers such as Teflon. While Teflon is not made from PFOA, the chemical is used as a critical emulsifying agent in the manufacturing process. Such large-scale manufacturing processes invariably lead to some environmental contamination, and unfortunately PFOA turned out to be environmentally very persistent. By the 1990s, it was clear that PFOA was showing up around the globe in amounts that could not be accounted for by release from the Teflon industry. Where was it coming from? Suspicion shifted to perfluorooctanyl sulphonate (PFOS), a chemical widely used in the production of stain-resistant materials for fabrics, carpets and furniture. When evidence accumulated that PFOS could indeed be converted to PFOA in the environment, the major manufacturer, 3M, decided to phase out production.

Surprisingly, the elimination of PFOS did not lead to a reduction of blood levels of PFOA. Clearly, there had to be another source of contamination. And there was strong motivation to find it, because by the 1990s researchers were demonstrating that such polyfluorinated compounds caused reduced birth size in rodents, developmental delays, alteration of fat metabolism and disruption of thyroid hormones. Of particular concern was the association of PFOA with liver, pancreatic and testicular cancer in animals.

Attention now shifted to some of the chemicals that had replaced PFOS, which were thought not to be capable of giving rise to PFOA. Could compounds referred to as fluorotelomers, or the greaseproofing polyfluoroalkyl phosphate esters, be the source? Scott Mabury's evidence points in that direction. His group dosed rats with various PAPs and found that they biotransformed into polyfluorinated carboxylic acids (PCFAs), with one particular PAP giving rise to PFOA. Calculations based on the rat data show that human blood levels of PFOA can be accounted for to a large extent by exposure to PAPs from food packaging material.

We are still left to puzzle over what this means. No link between PFOA and cancer has been found in humans, even with occupational exposures far exceeding the three to four parts per billion normally detected in the general population. A few studies have suggested an association between exposure in girls and a later onset of puberty, lower birth weight when mothers have higher blood levels of PFOA, and decreased semen quality in men, but most studies have not found such effects. Still, concern cannot be dispelled, because PFOA and related compounds are environmentally persistent.

Why should polyvinyl chloride (PVC) waste not be incinerated?

Heating plastics made of PVC to a high temperature can result in the formation of dioxins. And these are nasty compounds. Never made on purpose, dioxins are byproducts of some industrial processes, but can also form naturally when forests burn or volca-noes erupt. They rank among the most toxic substances known and have been linked with immune system depression, various can-cers and reproductive disorders. The term *dioxin* actually refers to any of roughly two hundred different compounds that have a simi-lar molecular structure. Their degree of toxicity varies, with the most worrisome being the ones that contain chlorine in their structure. All seventy-five known chlorinated dioxins are believed to be toxic to some organism at some level.

It should be stated, however, that most toxicity data is derived from animal studies, and that when it comes to dioxins, there is great variation even among closely related species. For example, tetrachlorodibenzodioxin, or TCDD, the most dangerous dioxin known, is thousands of times more toxic to guinea pigs than hamsters. But the measure of toxicity used in this case is the LD-50, the amount it takes to kill half of an animal population. Unfortunately, that doesn't tell us much about long-term exposure to trace amounts, which is what humans are concerned about. This is very difficult, if not impossible, to determine.

We can get some important clues from following the health status of workers who are most likely to be exposed to dioxins, from the consequences of a tragic accidental release of dioxins at a chemical plant in Seveso, Italy, in 1976, and from monitoring Vietnam veterans who were exposed to the defoliant Agent Orange, which was contaminated with dioxins. Following people who lived near toxic waste dumps that harboured dioxins, such as at Times Beach in Missouri, can also be revealing. While some data suggest increases in cancer and reproductive difficulties, there is much debate among experts about the long-term effects of exposure to trace amounts of dioxins. Be that as it may, there is no benefit to having these compounds around, so measures to reduce exposure are appropriate.

Since dioxins are notoriously persistent in the environment, they end up being distributed around the globe. Of great concern is that these are fat-soluble compounds, meaning that they can bioaccumulate. From the air, they settle on crops, which are eaten by animals, which we in turn eat. In the end, we become a storage receptacle for dioxins, possibly making us more prone to cancer, diabetes and reduced sperm counts. Eating less meat reduces exposure to dioxins, but of course, the best safety measure is to not produce them in the first place.

It turns out that the largest sources of dioxins are landfill fires and the barrel burning of household waste. Medical and municipal waste incinerators also contribute significantly, as do some chlorine-based bleaching processes in paper mills. Traces of dioxin are also produced when herbicides such as 2,4-D or the antibacterial agent hexachlorophene are manufactured, although industry has taken great measures to reduce any release to the environment. But incineration is a problem. All attempts should be made to keep PVC out of the incineration stream, and homeowners who burn garbage in their backyards have to be educated about the risks to public health that come from burning PVC plastics.

Even if measures are taken to reduce incineration of PVC, the polyvinyl chloride industry will always be plagued by dioxin problems. That's because dioxins can also form as byproducts when vinyl chloride, the *monomer* needed to make PVC, is manufactured from ethylene. Current technology has minimized this problem, but some environmentalists still clamour for the total elimination of PVC production. This is completely unrealistic given the extensive use of this plastic. You'll find it in a myriad of items from water pipes and floor tiles to artificial Christmas trees and wire insulation. Many flexible food wraps are made of PVC, and so are those hard-to-open rigid clamshell packages.

Although total elimination of PVC is unreasonable and unnecessary, reduction is possible. Many green-oriented companies are substituting other packaging material for PVC because of the incineration problem. One of the best examples is Microsoft, which phased out PVC packaging, eliminating hundreds of thousands of pounds of PVC waste. We can do our share as well by recycling as much as possible and never burning any plastic waste in our backyards.

In the 1970s, seventeen people who supposedly were involved in Irish Republican Army terrorist attacks on the British mainland were convicted on the basis of finding what chemical on their hands?

Nitroglycerin. The "Guildford Four," "Maguire Seven" and "Birmingham Six" were all convicted for their roles in bombings that resulted in multiple deaths in the towns of Guildford and Birmingham. Forensic evidence played a key role in the convictions, with tests demonstrating the presence of nitroglycerin, an explosive component, on the hands of the alleged bombers and bomb makers. Unfortunately, as it later turned out, the tests used were not as reliable as portrayed during the trial.

The Maguire Seven were convicted of making the bombs the Guildford Four used to blow up a pub. Major evidence was provided by a technique known as thin layer chromatography, which was used to detect nitroglycerin in swab samples taken from the hands of the accused. Sensitive sniffer tests found no evidence of nitroglycerin at the house where the bombs were supposed to have been assembled. A defence witness at the trial correctly stated that the thin-layer test was incapable of distinguishing between the handling of explosives and the handling of nitroglycerin tablets commonly used for the treatment of angina.

Furthermore, later findings revealed that ether, used as a solvent in the test kit, could itself have been contaminated with nitroglycerin since the kits were manufactured by the Royal Armament Research and Development Establishment, which apparently also manufactured explosives. The kit used was, however, capable of distinguishing between nitroglycerin and nitrocellulose, which was another concern.

Playing cards are sometimes coated with nitrocellulose, which an earlier test known as the Griess test could have confused with nitroglycerin. Yet that was the test used to convict the Birmingham Six. In the Griess test, hand swabs are treated with sodium hydroxide to release nitrite ions from nitroglycerin. Reaction with sulphuric acid and alpha-naphthylamine then yields a pink colour, indicative of the presence of nitrite. But significant handling of playing cards can lead to a positive reaction. So can sloppy urinating—urine contains nitrite that can trigger a positive Griess test. Because of questionable forensic evidence and allegations that confessions had been extracted by force, the supposed IRA terrorists were later pardoned, but by that time they had served their sentences.

In the 1950s, women's shoes featuring heels made of this material were all the rage. What was that material?

The heels were made of acrylic—or if you want to be really technical about it, polymethyl methacrylate. But the common name for this plastic, depending on which company manufactured it, was Lucite, Perspex or Plexiglas. Acrylic is clear and resembles glass, so shoes that sported such a heel looked unusual and appealing. In a classic scene from the movie *How to Marry a Millionaire*, Marilyn Monroe shows off her shapely legs as she sunbathes wearing Lucite shoes. Don't get too excited—she also wears a bathing suit.

By the 1950s acrylics were already, as it were, well heeled. As early as 1877 German chemists Fittig and Paul had shown that molecules of a simple compound, methyl methacrylate, a colourless liquid, could be joined together to make polymethyl methacrylate, a tough solid. But it wasn't until 1936 that commercial production began under the name Plexiglas, with both the process and name patented by the German chemist Otto Röhm. During the Second World War this novel plastic, being clear and stronger than glass, found all sorts of applications, ranging from submarine periscopes to fighter-plane canopies and gun turret enclosures on bombers. And that had an interesting spinoff. Airmen who got shards of Plexiglas in their eye from shattered airplane canopies fared better than those who were injured by glass splinters. Acrylics turned out to be more compatible with human tissue than glass and did not cause as much inflammation. This observation led to the use of acrylics in the first hard contact lenses.

Acrylic plastics also turned out to be ideal for dentures and found a use in composite dental fillings. Hockey was also a beneficiary; the protective Plexiglas around the rink was far better than netting or wire mesh. Acrylic paints also appeared, basically consisting of pigments and polymethyl methacrylate suspended in water, so there was no worry about solvent vapours. And then you have acrylic fingernails. Actually, you have two types of acrylic fingernails: the type made of acrylic and glued onto the fingernail with cyanoacrylate—yet another type of acrylic—and the gel type that is painted onto the fingernail and is hardened by exposure to ultraviolet light. That involves some interesting chemistry, with polymerization actually taking place on the finger.

This kind of a reaction is initiated by the formation of highly reactive molecular species known as free radicals. Once formed, a free radical adds to a molecule of methyl methacrylate, which then becomes very reactive and adds to another molecule to form a *dimer*, which then latches on to another monomer, and pretty soon all the small molecules are zipped into a long chain. You now have a hardened acrylic nail.

But why do you have to sit with your fingers under an ultraviolet lamp? Because it's the UV light that generates the initial free radicals that get everything started. Ultraviolet light is energetic enough to break chemical bonds, which is exactly what it does to a *photoinitiator* that is incorporated into the mix. Under the effect of UV, it breaks apart into free radicals. These then start the cascade of reactions, resulting in a polymer. Of course, ultraviolet light is energetic enough to break other chemical bonds as well, including ones in DNA. That's why excessive exposure to the sun causes skin cancer. And that brings up an interesting question: Is there a risk of cancer by exposing the skin on the hands to ultraviolet light while waiting for the acrylic gel to harden?

Some concern about this possibility was generated by two Texas dermatologists in a paper submitted to the journal *Archives of Dermatology* in 2009. They reported diagnosing skin cancer on the fingers of two women, aged fifty-five and forty-eight, both of whom had had previous exposure to ultraviolet nail lights. The first had a fifteen-year history of twice-monthly UV nail light exposure; the second had had about eight treatments in one year, but that was several years before the first cancer appeared. Such case reports are interesting, but they are not very meaningful statistically.

Ultraviolet light–cured acrylic nails have been popular for some twenty years, with millions of women using them. Any significant risk of skin cancer on the hands would have already been noted epidemiologically. Actually, UV exposure from nail lights is quite small in comparison to exposure from other UV sources such as sunlight. Living in Texas exposes one to significant UV radiation.

Calculations show that exposure from a nail lamp is equivalent to spending an extra minute and a half to three minutes a day in sunlight between salon visits, the time depending on whether the lamp has one or two bulbs. Basically, the two case reports do not make for a compelling case and should not cause panic. The time spent under the lamps just isn't long enough to present a significant risk. It is also interesting to note that since the paper originally appeared, there have been no further reports of skin cancers linked to nail lights.

One would have expected other dermatologists who read the paper to chime in with case histories, as is often the case after such publications. No further reports have been published. So, there's no need to fret about acrylic fingernails, at least not because of UV exposure. But in addition to the acrylic monomers and photoinitiators, there are cross-linking agents, reaction accelerators, plasticizers and pigments. So irritation and allergic reactions are always a possibility. Of course, Lucite heels can be worn safely. And they're still kicking around, in more shapes and styles than ever.

CHEMISTRY
UP CLOSE

What is the link between diamond, graphite and buckyballs?

They are all composed only of carbon. That sounds confusing, because these substances have dramatically different properties. But what determines the properties of any substance? Simple: the properties depend on the type and number of atoms they are composed of, and on how those atoms are joined together. Water and hydrogen peroxide, for example, are both composed only of hydrogen and oxygen, but a molecule of hydrogen peroxide has two hydrogen atoms and two oxygen atoms, while water has two hydrogens and one oxygen. So even though they have the same components, they have dramatically different properties.

Two substances may even contain exactly the same types and numbers of atoms, but still be totally different because of the way the atoms are joined together. A classic example is ammonium cyanate and urea, both of which correspond to the formula N_2H_4CO but are very different. Urea is a breakdown product of proteins and shows up in the urine, while ammonium cyanate is a mineral substance. These two hold a fundamental place in chemical history because of a classic experiment carried out by Friedrich Wöhler in 1828. Wöhler heated ammonium cyanate and converted it into urea. At the time, it was believed that substances found in living systems, such as urea, were held together by some sort of "vital force" that could never be reproduced in the laboratory. Wöhler's experiment disproved this and spawned the modern field of organic chemistry.

Before Wöhler, organic chemistry was described as the chemistry of substances found in living or once-living systems—substances believed to be beyond the scope of laboratory synthesis since they could not be infused with the necessary vital force. Wöhler, however, had shown that organic molecules were not characterized by any mysterious force; their common feature was that they all contained carbon atoms. And these carbon atoms readily joined to each other as well as to other atoms in myriad ways, giving rise to a vast array of organic compounds ranging from fats and proteins to the flavour of vanilla and the odour of a rose. Diamond, graphite and buckyballs are examples of the variety of ways carbon atoms can join together, and are particularly interesting because they contain *only* carbon atoms.

In diamond, the atoms are arranged in a three-dimensional lattice, while in graphite they are joined together in an array of planar six-membered rings, arranged in layers. The diamond is very hard, while the possible movement of the layers in graphite makes it into an excellent lubricant. *Buckyball* is a colloquial term for a molecule in which sixty carbon atoms are joined in a sphere, as if located at the corners of the panels used to make a soccer ball. The curious name comes from the geodesic dome concept pioneered by architect Buckminster Fuller. In 1996, Harold Kroto, Richard Smalley and Robert Curl received the Nobel Prize in chemistry for their synthesis of fullerenes, of which buckyballs are an example. Despite the fact that they are composed of the same atoms, diamond, graphite and bucky-balls have completely different properties. Such is the magic of molecular bonding.

What element can be used both as an abrasive and as a lubricant?

Carbon. Both graphite and diamond are naturally occurring forms of the element carbon. Diamond is the hardest substance that exists, and it readily scratches any surface. Graphite, on the other hand, allows surfaces to readily slide over each other. How can two substances, both composed of carbon, be so dramatically different? As we just saw above, it is all a matter of how the carbon atoms are joined together. In diamond, each carbon atom is bonded to four others, forming a three-dimensional lattice. In graphite, each carbon is joined only to three others, forming a network of six-membered rings arranged in planar sheets. These planar layers are stacked one on top of the other and can slide over each other, hence the lubricating effect.

Actually, it is a film of water vapour trapped between the layers of carbon atoms in graphite that allows for the easy movement of the layers. While graphite is a great lubricant for locks, machine parts and motors here on earth, it cannot be used in a vacuum, where there is no moisture to be trapped. Therefore, graphite cannot be used to lubricate moving parts of satellites or space vehicles.

Where does graphite come from? Most of it is mined. The first large deposit to be discovered was in England, back in 1565. There wasn't much use for the material then, except for marking sheep. But it wasn't long before it became apparent that this substance would also leave a mark on paper, and the first pencils made of sticks of graphite wrapped in string appeared. At the time, the black substance was thought to be a form of lead and was given the name "plumbago," from the Latin for "lead ore." It seemed logical, since lead itself had long been used as a writing instrument. The metal is soft and leaves a thin layer behind when rubbed on a surface.

But graphite isn't lead, of course; however, the name has stuck when referring to a pencil. By the seventeenth century, sticks of graphite were being inserted into holes bored in pieces of wood, and the pencil industry was born. The most important use for graphite at the time, however, was to line moulds used to make cannon balls. This made the graphite deposits so important that control was taken over by the Crown and graphite for pencils actually had to be smuggled out.

Today, synthetic graphite is also available, made by heating various carbon-containing substances, such as coke, to extremely high temperatures. Fibres of synthetic graphite can be produced by heating rayon or polyacrylonitrile fibres to drive off all the hydrogen, oxygen and nitrogen they contain, leaving only carbon behind. When combined with synthetic resins, these yield strong, lightweight materials used for aircraft and automobile parts, golf clubs and tennis racquets.

Perhaps the most famous carbon fibre was created by Thomas Edison in 1879. To make his early light bulb filaments, Edison formed cotton threads or bamboo slivers into the proper size and shape and then baked them at high temperatures. Cotton and bamboo consist mostly of cellulose, a natural linear polymer made of repeating units of glucose. When heated, the filament was *carbonized*, becoming a true carbon copy of the starting material. Tungsten wire soon displaced these carbon filaments, but they were still used on U.S. Navy ships as late as 1960 because they withstood ship vibrations better than tungsten.

Graphite can also be converted to diamond using high pressures and temperatures. The diamonds produced in this fashion are small and are used for industrial purposes, such as in diamond-tipped drill bits and saws, as well as in abrasive powders to polish items like marble and gemstones. Gem-quality diamonds are not made synthetically, unless one considers formation deep in the ground from carbonaceous matter over billions of years a synthetic method. Natural diamonds make it to the surface through deep volcanic eruptions. The largest ever found was the Cullinan diamond, discovered in a South African mine in 1905. It weighed 621 grams and was eventually cut into nine smaller diamonds, the two largest of which are among the Crown jewels of England.

The world's first commercially available laundry powder was Persil, introduced by the German company Henkel in 1907. Where did the name Persil come from?

"Persil" derives from perborate and silicate, two key components in the product. It was introduced as an improvement over the action of soap, the traditional cleaning agent first formulated around 1500 BC. Just heat some sort of fat with ashes from a wood fire and you get soap. The ashes supply the alkaline chemicals needed to break down the molecules of fat and convert them into salts of fatty acids, which we know as soap. One end of the soap molecule has an affinity for water, the other for oily substances. Washing with soapy water, then, removes oily residues from a surface.

While soap cleans well by emulsifying and removing greasy stains, it does present some problems. It isn't great on coloured stains, and it forms a precipitate when used in water that has a high mineral content. This "scum" is hard to rinse away and it dulls clothes. Persil addressed both of these problems.

Sodium perborate is an oxygen-releasing agent, and oxygen is effective at destroying stains. As the prototype oxidizing agent, it can steal electrons from molecules. Since electrons are the glue that holds molecules together, exposure to oxygen can break down complex molecules, such as the ones responsible for stains. This is why laundry was traditionally either hung out to dry or spread out over grassy fields. Not only did this expose the fabric to oxygen, but also to ultraviolet light from the sun, which can also break down coloured molecules. Sodium perborate did the work of the air and the sun at the same time.

The addition of sodium silicate had a water-softening effect, meaning that minerals like calcium and magnesium responsible for forming a scum with soap were, in a sense, neutralized. These minerals react with silicates to form precipitates, just as they do with soap, but the difference is that these precipitates are readily rinsed away and tend not to deposit on the fibres of the cloth being washed. Silicates have great suspending and anti-redeposition qualities. Today's detergents are chemically far more complex than the original Persil, and Persil itself has a range of products to cater to different needs, but it will always retain its place in history as the "first self-acting laundry detergent," and the image of the White Lady introduced in 1922 and featured on numerous placards and signs remains an advertising classic.

What comes next in the following sequence: milli, micro, nano, pico?

Femto. These terms are known as "power prefixes" and are used to denote small numbers. Each one is one thousand times smaller than the previous notation. A millimetre, for example, is one thousandth of a metre. A microlitre is one millionth of a litre. A nanometre is one billionth of a metre, and a picogram is one thousandth of a billionth of a gram. The next one in line is femto, derived from the Danish word *femten*, meaning "fifteen." A femtosecond would be ten to the power of negative 15 of a second. A femtogram would be 10^{-15} of a gram, or one quadrillionth of a gram.

These are incredibly small numbers, but amazingly, techniques are now becoming available to detect substances down to femto levels. A technique that couples atomic force microscopy with Fourier transform infrared spectroscopy allows analysis of substances at the femtoscale level. At that level, we will find that everything is contaminated by everything else!

What naturally occurring chemical is the best candidate for producing "green gasoline"?

Cellulose. The term *green gasoline* refers to a mixture of compounds that is chemically very similar to gasoline sourced from petroleum, but originates from plant material. It is not ethanol or biodiesel, both of which can also be produced from plants. Ethanol is made by fermenting plant sugars, and biodiesel is made by treating animal or vegetable fats with methanol. Neither of these is chemically equivalent to gasoline. But experiments are under way to process cellulose, the main structural material of plants, into a fuel that would, in essence, be identical to gasoline.

Cellulose is composed of long chains of glucose molecules. When heated to a temperature of roughly 500 degrees Celsius, the chains break apart to produce small molecules called *anhydro sugars*. These are the molecules that can be converted to gasoline. But here is the problem: the anhydro sugars contain a lot of oxygen atoms, a problem in terms of combustion. Gasoline is a mixture of molecules called hydrocarbons, so called for the simple reason that they are composed only of hydrogen and carbon atoms joined together.

When hydrocarbons undergo combustion, the molecules are torn apart and the hydrogen atoms and carbon atoms both react with oxygen. Hydrogen in the hydrocarbons converts to water, and carbon to carbon dioxide. The energy released in this process vapourizes the water into water vapour. Both the water vapour and the carbon dioxide then expand dramatically as the temperature increases. It is the expansion of these gases in an engine that moves the piston.

To make green gasoline, the anhydro sugars produced from cellulose have to be stripped of their oxygen content, leaving a hydrocarbon framework that can react with oxygen during combustion. This oxygen stripping is carried out using a naturally occurring mineral called zeolite. Actually, there are many zeolites, all characterized by having a basic structure composed of alumino-silicates. Zeolites are very porous minerals and are permeated with microscopic channels into which molecules can enter.

Different zeolites have different pore sizes and can play host to different-sized molecules. Some zeolites, for example, have just the right size channels to allow calcium or magnesium ions to enter, and therefore are effective water softeners. Others can absorb organic molecules that are responsible for smells and can act as odour neutralizers. And then there are zeolites that have pores of just the right size to allow anhydro sugars to enter. When these sugars squeeze into the pores, they undergo a chemical reaction whereby they lose oxygen and convert to hydrocarbons that can be used directly as gasoline. Essentially the surface of the zeolite serves as a catalyst for the decomposition of anhydro sugars to hydrocarbons.

So far, production of green gasoline is only at the laboratory stage. But the potential is great. There is plenty of raw material available in corn stalks, sawdust and non-food crops such as switchgrass that can be grown without fertilizer. We eventually will run out of petroleum, but the raw materials for green gasoline are renewable. And there is an added bonus: when the crops grow, they use up carbon dioxide from the air, countering some of the carbon dioxide released during combustion. There is a good chance that, in the not-too-distant future, we will be driving around powered by plants.

Why is Fix-A-Flat looking for an alternative to 1,1,1,2-tetrafluoroethane, its tire-inflating ingredient?

Also known as (HFC)-134a, tetrafluoroethane inflates a tire very effectively, but it also inflates the concentration of global-warming gases in the atmosphere. The use of this gas is going to be curbed, and it will have to be replaced by a substance with a lower global-warming potential as environmental regulations become more stringent. There's no question that pumping up a flat tire with Fix-A-Flat is a lot easier than changing the tire. It's a very clever product.

Fix-A-Flat dispenses a *polyolefin elastomer*, a chemical very similar to the rubber of which the tire is made. The specific polymer used is a trade secret, but it is probably something like polyisobutylene or ethylene propylene rubber, either of which can be dispensed as a foam that then hardens to fill holes in the tire from the inside. This is actually the same technology as used to make Crocs shoes. A foam is just a suspension of gas bubbles in a liquid, and the gas used to create the foam in Fix-A-Flat currently is tetrafluoroethane. The advantage of this compound is that it is easily liquefied under pressure and is inert, meaning that it will not react with other components in the product. When released from the can, it immediately evaporates, dispensing the polyolefin as a foam and simultaneously inflating the tire.

Some of the early versions of such products used dimethyl ether, propane or butane as propellants. These are highly flammable gases, and in some cases they caused explosions when the inflated tires were subsequently taken to be repaired. Using a metal tool to fix the leak on steel-belted tires sometimes caused a spark and set off a blast. Mechanics were not happy. With tetrafluoroethane there is no such problem, but the gas does eventually find its way to the atmosphere, where it contributes to global warming. That's because it absorbs some of the infrared light that is being radiated from the earth. When molecules absorb infrared radiation, they are energized and move about faster. That increase in molecular motion is what we sense as heat.

The global-warming potential of a gas depends on its lifetime in the atmosphere, its ability to absorb infrared radiation, and the specific wavelengths of infrared light it absorbs. Even if a gas absorbs a lot of infrared light, it still may not have much of a warming effect if other components of the atmosphere already absorb most radiation at that wavelength. But tetrafluoroethane absorbs infrared light at wavelengths that would otherwise pass through the atmosphere. A "global-warming scale" has been developed to estimate the contribution of various greenhouse gases to global warming. The effects are measured relative to carbon dioxide, which has been assigned the value of I.

On this scale, tetrafluoroethane, if present in the atmosphere for a hundred years, would have a global-warming potential of I,300. The next generation of gases used in products such as Fix-A-Flat will have to have global-warming potentials under I50. Currently, the best candidate is trans-I,3,3,3-tetrafluoro-propene, which comes on the scene with a low global-warming potential of about 6. Its use in Fix-A-Flat products would be a chemical triumph, although a minor one since the contribution of such products to global warming in comparison to other greenhouse gases is minute.

What common air pollutant has a climate-cooling effect?

Sulphur dioxide. We hear a great deal about gases in the atmosphere that can cause global warming. Carbon dioxide is the most famous one, but ozone, methane, freons and nitrous oxide are also important contributors. By contrast, we don't hear much about global cooling. Yet this is also an important phenomenon. Indeed, cutting sulphur dioxide emissions, a major goal of industrialized nations, is actually accelerating global warming. Coal and oil always contain small amounts of sulphur, which on combustion reacts with oxygen to yield sulphur dioxide. This gas then reacts further with oxygen to form sulphur trioxide, which in turn forms sulphuric acid or reaction with moisture in the atmosphere. Sulphuric acid aerosols, basically tiny droplets of sulphuric acid, reflect sunlight, preventing it from heating up the earth's surface.

Not all the sulphur dioxide in the atmosphere is due to human activity; volcanoes spew out huge amounts of the gas. Back in 1783, an eight-month-long eruption in Iceland resulted in a series of unusually cold winters in the northern hemisphere, even in America. The winter of 1784 was so cold that the Mississippi actually froze as far south as New Orleans. There is no doubt that, since the Industrial Revolution, sulphur dioxide emissions caused by human activity rose significantly, but the cooling effect was more than cancelled out by warming due to vast amounts of carbon dioxide and other greenhouse gases produced. As the West, and increasingly China and India, reduce sulphur dioxide emissions, global warming is set to become more significant unless other measures are introduced. Relaxing efforts to cut back on the release of sulphur dioxide is not an option. Sulphuric acid aerosols cause acid rain, which is devastating to ecosystems, and inhalation of sulphuric acid droplets is linked to bronchitis and asthma. We have to look at other means to reduce warming.

Carbon dioxide emissions are an obvious target, but attention also has to be paid to *black carbon*, essentially fine particles of soot. This is released into the atmosphere when fuels burn, especially diesel, coal, firewood and dung. The fine particulate matter has a "global dimming" effect, blocking some of the sunlight from reaching the earth. But unlike sulphuric acid aerosols, the particles don't reflect the light, they absorb it and heat up. This in turn warms the atmosphere. Another problem arises when the particles settle on snow or ice, absorbing light that would otherwise be reflected, causing melting.

Black carbon can even cause climatic problems. By absorbing sunlight over the oceans, it prevents the water below from warming up and evaporating, which in turn reduces cloud formation, which then changes the climate. Reducing cooking with firewood and dung in Asia would be an important step in reducing black carbon in the atmosphere, but that is not likely to happen in the short term. But filtering emissions from diesel engines is certainly technically feasible and is becoming even more important as we cut sulphur dioxide emissions and lose some of the global cooling they provide.

✣

Alkyl polyglycosides are surfactants that are being
increasingly used in "green" cleaning products.
What makes them green?

They are synthesized from cornstarch and coconut or palm oil,
which are "sustainable" resources. Surfactants—the term deriving
from "surface active agents"—are the key to cleaning. They disperse grease, improve the wetting of surfaces, lubricate, emulsify,
neutralize static and, depending on which specific surfactant is
used, can either cause or eliminate foaming.

Broadly defined, surfactants are compounds that, when dissolved in water, concentrate at the surface. The most familiar of these
is soap, which produces foam due to its action at the water—air
interface and removes grease because of its activity at the water—
oil interface. Without a doubt, soap, made from animal fat and
wood ashes, was the first manmade surfactant, but it was not the
first surfactant used. That honour likely belongs to the soapwort
plant, which produces compounds called saponins. Boiling
soapwort root produces a solution that foams and has cleaning
properties. Saponins are compounds made in the plant by a
combination of glucose with steroids or molecules called terpenes.
The glucose end of the molecule is attracted to water, while the
steroid or terpene has an affinity for oil, so the molecules
concentrate at the water—oil interface. Essentially, a link is formed
between the oil and the water, allowing the oil to be rinsed away
from a surface.

Saponins have a gentle cleaning action and are still sometimes
used for very delicate fabrics. But when synthetic chemistry came
to the fore in the twentieth century, soaps and saponins were
replaced by the more effective synthetic detergents, such as the
alkylbenzene sulphonates, which were derived from petroleum,
a non-renewable resource. Green products today are based on
sustainability and tend to feature "natural" ingredients. Alkyl

polyglycosides fit this requirement. The water-soluble end—units of glucose linked together—comes from cornstarch, while the attached oil-soluble end is made from alcohols derived from coconut or palm oil. There is still some synthetic chemistry involved in making the desired alcohols from palm or coconut oil and joining these to the cornstarch derivative, but this is a more environmentally friendly process than making totally synthetic surfactants.

Not every manufacturer is jumping on the "natural" bandwagon, for the simple reason that natural is not always better. While the alkyl polyglycosides have some environmental advantages, designing synthetic surfactants and combining them with protein, fat and carbohydrate-digesting enzymes is a better way to go. This can allow for washing in cold water, which has tremendous benefits. About 85 per cent of energy used for laundry comes from heating the water, and about 3 per cent of household energy use goes to heating water for laundry. Developing a cold-water laundry product has far greater benefits than substituting so-called natural surfactants for synthetics.

Professor Galen Suppes of the University of Missouri received an award for green chemistry for developing a use for glycerin (glycerol). What industry produces a massive amount of glycerin and will benefit from this technology?

The biodiesel industry. Biodiesel can be made either from vegetable oils or from animal fat by reaction with methanol. It is commonly referred to as a "green" fuel because, rather than being sourced from petroleum, it is made from renewable resources. But for every nine kilograms of biodiesel produced, about one kilogram of crude glycerol byproduct is formed. The economics of biodiesel production therefore depend heavily on finding uses for this glycerin byproduct. If a high enough demand for glycerin can be found, the cost of biodiesel may be reduced by as much as ten cents a litre. One possibility is to convert glycerol into propylene glycol, a chemical that has a great many potential uses. It can serve as an alternative to toxic ethylene glycol in antifreeze, and it can be used in the production of polyesters, pharmaceuticals, cosmetics, detergents, paints, flavours and fragrances.

Right now, most propylene glycol is made from propylene, which in turn is derived from petroleum or natural gas. It cannot be produced as cheaply as ethylene glycol, so antifreeze based on this safer chemical is more expensive. Finding a way to convert glycerin generated by the biodiesel industry to propylene glycol would serve not only to reduce the cost of biodiesel, but also to reduce the cost of propylene glycol antifreeze, both of which are definitely "green" developments. Several routes to the preparation of propylene glycol from renewable feedstocks have been explored. Treating sugars with hydrogen at a high temperature using certain metal catalysts can be carried out, but using excess glycerin from biodiesel production would be preferable. This is where Professor Suppes made his contribution. He and his colleagues developed a method of reacting glycerin with hydrogen using a new copper-chromite catalyst at readily achievable temperatures and pressures. If this technology can be scaled up, we will be looking at cheaper biodiesel fuel and cheaper, safer antifreeze. The world will be a little bit greener.

Humans and animals exhale carbon dioxide with every breath. Why is this not considered to be a problem as far as global warming goes?

The carbon dioxide we exhale does not contribute to global warming for the simple reason that we also take up an equivalent amount of carbon dioxide from the air, albeit indirectly. Everything we eat can be traced back to photosynthesis, the process by which plants take up carbon dioxide from the air and use it to produce the vast array of organic compounds needed for life. Our bodies can be regarded as living engines that require fuel and oxygen to produce the energy needed to sustain life. In that sense, we are not all that different from a car. Both for us and for the car, the source of oxygen is the air, roughly 20 per cent of which is made up of oxygen. An internal combustion engine burns gasoline and spews out water, carbon dioxide and a few combustion byproducts. We, instead of gasoline, burn the carbohydrates, fats and proteins in food. Like gasoline, these organic compounds are converted to carbon dioxide and water, which we then exhale.

How is it, then, that we don't worry about the massive amounts of carbon dioxide that are released with every breath taken by the billions and billions of people and animals that inhabit the world? Because every atom of carbon in the exhaled carbon dioxide comes from food that was recently produced by photosynthesis. Everything we eat, save for a few inorganic components like salt, was in some way produced by photosynthesis. This is obvious when we eat plant products such as grains, fruits and vegetables, but of course it is also the case for meat. The animals we eat were raised on plant products. Indeed, a growing animal is basically a machine that converts plants into flesh. So, since all the carbon dioxide we exhale originated in carbon dioxide captured by plants during photosynthesis, we are not disturbing the carbon dioxide content of the atmosphere by breathing.

On the other hand, when we burn fossil fuels such as gasoline, we are releasing carbon dioxide that forms from carbon atoms that had been removed from the atmosphere millions and millions of years ago by photosynthesis and had then been sequestered in the coal, petroleum and natural gas that forms when plants and animals die and decay. By burning these commodities, we are increasing the current levels of carbon dioxide. Clearly, then, by living and breathing we are not contributing to global warming through the release of carbon dioxide. But can we help reduce global warming by dying? Probably. We no longer exhale carbon dioxide and it will be a long time before the carbon atoms in our body eventually make it back to the atmosphere as carbon dioxide. Of course, there are always plenty of new babies who start to respire as we expire.

INDEX

acetic acid, 96–97, 113–14, 115

acetone, 115, 231

acetone peroxide, ix–x

acetylcholine receptors, 120

acetylsalicylic acid, 121–22

acid rain, 268

acne, 176

Acokanthera schimperi (tree), 49

acrylic, 7, 246–49

Adams, Samuel Hopkins, 9–10

African crested rats, 48–49

African mango, 147–49

Agent Orange, 243

airplanes, oxygen masks in, 30

Ajinomoto (food company), 162

alchemy, 36–37, 227

alcohol

    in Coca-Cola, 10, 11

    consumption with guaranta,

153–54

    metabolism of, 113–14

alkaline diets, 122–23

alkyl polyglycosides, 269–70

aluminum, 239

Amazon basin, 152

amber, 77–80

American Medicine, 210

amethyst, 74–76

ammonium phosphate, 234

ammonium sulphide, 39–40

anaerobic bacteria, 103–4

androstencne, x–xi

anemia, 117–18

anhydro sugars, 263–64

animal husbandry, 108–9

Anita (spider), 62–63

Annual Review of Progress in Chemistry, 16

anthocyanins, 147

anticoagulants, 170

aphrodisiacs, 54, 107–9, 145

Apollo 13, 38

apoptosis, ix

Appert, Nicholas, 142

aqua regia, 61

Aqua Tofana, 108

Arabella (spider), 62–63

Archives of Dermatology, 248–49

argyria, 225–26

Artemis (Greek goddess), 74–75

arthritis, 131–32

artificial sweeteners, 156–57

asparagus, 145–47

Aspergillus terreus, 112

Aspirin, 121–22

asthma, 46, 237, 268

atoms, bonding, 253–55

atorvastatin, 112

Atwater, Wilbur Olin, 138

Augustus Caesar, 107

baby shampoo, 178–80

bacteria

anaerobic, 103–4

carried by cockroaches, 46

oral, 103–4, 156

production of nitrites,
133–34

production of vitamin B$_{12}$,
116, 118

silver as antimicrobial agent,
223–25 as source of
transglutaminase ("meat
glue"), 162

in uncooked meat, 163–64

and weight control, 139–40

baking powder, 141

baking soda, 141

balloons, 196–98

batteries, 61, 228

bauxite, 239

Bayer, Karl Josef, 239

beer, formation of head, 27

Bell, Chichester, 236

benzene, 188

benzopyrene, 102

Berliner, Emile, 236

Berthelot, Marcellin, 187

beta-carotene, 155, 176–77

BioAmber Inc., 76–77

biodiesel, 182, 263, 270–71

Bioelectric Shield, 76

biofilms, 103–4

"bioidentical hormones," 71–72

biological warfare, 56–58

bird's nest soup, 134–37

Birmingham Six, 245–46

bisphenol A (BPA), 213–21

Black, Joseph, 18–19

black carbon, 268

Blattella germanica (cockroach), 45

blister beetle extract, 107–9

blood

measuring alcohol
concentration, 114–16

pH level, 121–22

Blumenthal, Heston, 162

bobcat urine, 52

bombs

"fart," 41

phosphorus, 36

riot-control, 41

stink, 39–41
Bonaparte, Napoleon, 141, 206
boric acid, 194
Brand, Hennig, 36–37
Braun, Joe, 220
Brazilian hair-smoothing
    products, 180–81
Brazilian pawpaw, 89–92
breast cancer, ix, 71
Breathalyzers, 113–16
broccoli, 168–71
brominated vegetable oil,
    149–51
bromism, 150–51
Brown-Séquard, Charles-
    Édouard, 12–14
Brugnatelli, Luigi, 61, 228
buckyballs, 253–55

C. elegans (roundworm), ix
"cabbage goitre," 105
cadaverine, 103–4
caffeine, 152–53
    effect on spiders, 62
    in soft drinks and energy
        drinks, 9–12
calcium carbonate, 95–97
calcium hydroxide, 95–97
calcium oxide, 97
calcium polysulphide, 41
calories, measurement of,
    138–40
calorimeter, 138
Campaign for Safe Cosmetics,
    178–81
cancer
    "alternative" treatments, 71–72

breast, ix, 71
    linked to smoke and creosote,
        157–58
    and nitrites, 133
    prostate, 171
    purported cures, 90–92,
        145–46
    skin, 158, 176–77, 248–49
    treatment of, 32
Cancer News Journal, 145–46
Candler, Asa, 10–11
cannabis, effect on spiders, 62
cantharidin, 107–9
capture myopathy, 203
caramel colouring, 159–61
caramelization, 159–60
carbohydrates, 233
carbon, 253–57
    black, 268
carbon dioxide
    in baking, 140–41
    in beer, 27–28
    discovery, 18
    exhaled by humans, 272–73
    reaction with lime water,
        95–97
carbon monoxide, 60
carbonic acid, 122, 194
carbonyl sulphide, 195
cardiac glycosides, 48–49
Carissa spinarum (plant), 87–89
carnauba wax, 235–36
carotenoids, 147, 155
castoreum, 86
celebrities, pseudo-science
    endorsed by, 71–74
celery, 133–34

cellulose, 256, 263–65
Center for Science in the Public
    Interest, 160
Centers for Disease Control,
    215–16
charcoal, use in explosives
    manufacture, 20
Charles II (king of England), 6
chemical warfare, 56–58
chemistry
    "green," 76–77
    organic, 254
    polymer, 187
    teaching of, 18–19
Chernobyl (Ukraine), 92–93
chewing gum, 156
chimney sweeps, 158
China
    mythology, 58–60
    traditional Chinese medicine,
        54–56, 136, 164–66
chocolate, phenylethylamine in,
    53
cholesterol, 110–12, 143–44,
    148, 165–66, 189
chromatography, 15–18
    gas, 17
    paper, 16
    thin-layer, 245–46
chuno (potato product), 22
cinnabar, 22, 227
citric acid, 222, 230
Coca-Cola, 9–11
cocaine, 4, 10–11
cockroaches, 45–46
Colgate-Palmolive Co., 182
colloidal silver, 225–26

combustion, 29–30
    of hydrocarbons, 195,
        263–64
    smoke produced by, 58–60
    and sparklers, 198–99
Conan Doyle, Sir Arthur, 200
concrete, self-cleaning, 31–32
copper
    antimicrobial properties, 224
    and oxidation of alcohol,
        113–14
cosmeceuticals, 175–78
cosmetics, 175–78, 178–81
Cremer, Erika, 17
creosote, 158
cretinism, 105, 106
cruciferous vegetables, 168–71
Cruise, Tom, 73
cubilose, 135–37
Cullen, William, 18
Cullinan diamond, 257
cuppelation, 20–22
cured meats, 132–34
Curl, Robert, 254
custard apple, 89–92
cyanoacrylate visualization,
    33–34
cyanocabalamin, 118
cyclosporine, 103
cytochrome p450 enzymes,
    101–3

Daly, John, 119–20
Dendrobates azureus (frog), 47
deodorizers
    para-dichlorobenzene, viii–ix
    testing of, 41

Derby, Elias, ix
dermatitis, 176
detergents, 184, 257–58
  phosphates in, 37
dextrin 198–99
dextrose, 106
diabetes 148
diamonds, 253–55, 257
dietary supplements, 147–49,
  170, 176, 190, 225–26
digitalis 48
digoxin, 189
dimercaprol, 229
dimethicone, 184–85
Dionysus (Greek god), 74–75
dioxins, 242–44
Diprivan 123–25
Drake, Colonel Edwin, 201, 202
drunkenness
  measuring with
    Breathalyzers, 113–16
  purported antidotes, 74–75
Du Pont (chemical company),
  20
duck cotton, 39
duct tape, 38–39
Dufresne, Wyle, 161, 162
Durand, Peter, 142
dyes
  manufacture of, 23–24
  synthetic, 24, 67–69

Ecuadorian tree frogs 47–48,
  119–20
Edison, Thomas, 235–36, 256
electromagnetic radiation, 76,
  92–93, 115

electroplating, 61–62, 228
Elkington, George, 228
Elkington, Henry, 228
emulsions, 185
Endo, Akira, 110–11, 112–13
energy bracelets, 70
energy drinks, 12, 153
epibatidine, 47–48, 119–21
Epipedobates tricolor (frog), 47–48,
  119–21
Epsom salt, 193–94
Eschenmoser, Albert, 116
esnesnon, 69
espionage, 34–35
ethanol, 113–14, 263
ether, 246
ethylene, 188
Eugénie (empress of France), 24
European Food Safety
  Authority (EFSA), 161,
  217
explosives, ix–x, 16, 20, 205,
  230–32

faith healing, 87–89
Faraday, Michael, 196–97
"fart bombs," 41
Fat Duck (restaurant), 162
fats, saturated, 183
Ferrosan (pharmaceutical
  company), 177
ferrous sulphate, 222
fertilizers
  manufacture of, 205–6
  phosphates in, 37
ferulic acid, 146
fetal alcohol syndrome, 129

Fielding, Henry, 130
fingernails, acrylic, 247–49
fingerprints, detection, 33–34
fire extinguishers, 234–35
Fittig and Paul (chemists), 247
Fix-a-Flat, 265–67
fixatives, 186–87
Fleming, Alexander, 110
fluitex therapy, 69
fluorotelomers, 242
food
    additives, 84–87, 104–6,
        132–34, 143–44, 149–51
    canning, 141–43
    colourings, 159–61, 164–66
    flavourings, 157–59
    metabolism of, 138
    packaging, 240–41
Food and Drug Administration
    (U.S.), 46, 165–66,
    215–16, 217
Food Revolution (TV series), 84
forensic science, 33–34,
    199–201
formaldehyde, 178–81, 231
free radicals
    and cancer treatment, 32
    and polymerization, 247–48
    scavengers of, 88, 147
    and self-cleaning glass, 31
    skin damage caused by, 176
french fries, 139
frogs, "poison dart," 47–48,
    119–20
fruit flies, 73–74
Fuller, Buckminster, 254
fullerenes, 254

furanocoumarins, 102
fusarium, 56, 58

Gaedeke, Friedrich, 4
Galvani, Luigi, 61
garbage, incineration of, 242,
    244
gas chromatography, 17
"geopathic stress," 82
German Society of Toxicology,
    217
gilder's palsy, 226–28
gin, 129–32
Gin Act (Britain), 130
Gin Lane (Hogarth), 130
glass
    self-cleaning, 30–32
    tempered, 4–7
glucoraphanin, 168–69
glucoronic acid, 102
glutathione, 146
glycerin, 270–71
Godfrey, Ambrose, 234–35
goitres, 105
gold, smelting of, 227–28
gold-plating, 61–62, 228
Gonzalez, Nicholas, 71–72
Goodyear, Charles, 197
Goppelsröeder, Friedrich, 16–18
Gourbin, Emile, 200
graphite, 253–57
graviola, 89–92
Gray, Derek, 224–25
Great Wall of China, 59
"green chemistry," 76–77
"green gasoline," 263–65
greenhouse gases

carbon dioxide exhaled by humans, 272–73
and farming, 167–68
"global-warming scale," 266
methane, 203
in tire-repair products, 265–67
Greenpeace, 183
Griess test, 246
guarana, 12, 152–54
Guildford Four, 245–46
Guinness (stout), 27–29
Gummi bears, 29
gunpowder, 16, 20, 205

Haber process, 206
Haig, Alexander, 56
Hancock, Thomas, 197
Harvard School of Public Health, 138–39
Harvey, Paul, 132
Hashimoto's disease, 105
hatter's shakes, 226
Health Canada, 165–66, 214
Health Sciences Institute, 90–91
heart disease
congestive heart failure, 48–49
and elevated cholesterol, 110–12, 143–44
and oral bacteria, 103
and palm oil, 183
and red yeast rice, 165
and saponins, 189–90
herbal remedies, 87–89
hexamethylene triperoxide diamine (HMTL), 231

Hippocrates, 79
histones, 146
hockey rinks, protective glass around rinks, 5, 7
Hofmann, August Wilhelm von, 187
Hogarth, William, 130
Holmes, Sherlock (fictional character), 200
homeopathic remedies, 73, 109
hormone therapy, 13–14
hormones, 68–69
"bioidentical," 71–72
"imbalanced," 72
leptin and adinopectin, 149
testosterone, 14, 206–8
houseflies, 50–51
"human magnets," 92–95
hydrocarbons
combustion of, 263–64
polycyclic aromatic (PAHs), 158–59
hydrodesulphurization, 195
hydrogen, in balloons, 196, 197
hydrogen peroxide, 230, 231
hydrogen sulphide, 39–40, 195
hyperventilation, 121–22
hypodermic syringes, 3–4
hypothyroidism, 105

Imedeen (skincare product), 177
"in vitro" meat, 167–68
infrared clothing, 70–71
inks
addition of pectin to, 8
gall, 221–22
invisible, 34–35

insects
  blister beetle, 107–9
  cockroaches, 45–46
  fruit flies, 73–74
  houseflies, 50–51
  monarch butterfly, 48
intrinsic factor, 117
invisible ink, 34–35
iodine, 33, 104–6
Irish Republican Army (IRA),
    245–46
iron oxides, 239–40
"iron-gall ink," 221–22
Irvingia gabonensis, 147–49
isothiocyanates, 169

Jackson, Michael, 123–24
Jacobi-Kleemann, Margrit, 64
James Randi Educational
    Foundation, 93–95
Jeffery, Elizabeth, 170
jewellery
  amber necklaces, 79
  energy bracelets, 70
  "metaphysical," 75–76
Job, Oswald, 34–35
John Waddington Ltd., 8–9
Johnson, B.J., 182
Jones, Stan, 225–26
Joseph Andrews (Fielding), 130
juniper berries, 130–31

kangaroo meat, 203–4
kangatarians, 203–4
keratin, 55
kerosene, 202
Khan, Ahmed Ali, 230, 231

Kier, Samuel, 201–2
Kingdon, Jonathan, 49
Kissinger, Henry, 166
Koller, Karl, 4
Krazy Glue, 33–34
Krebs cycle, 79
Kroto, Harold, 254

Latelle, Marie, 200
laughing gas, 197
Lawrence Berkeley National
    Laboratory, 38
L-cysteine, 85
lead
  bullets made of, 195, 196
  in silver ore, 20–21
lead nitrate, 34–35
leeches, 73
Leimadophis epinephelus (snake), 47
Letterman, David, 86
Li Zheng-yu, 58–60
Liberles, Stephen, 52
light bulb filaments, 256–57
lime water, 95–97
Lipids in Health and Disease, 148
Lipitor, 112
Liquiambar orientalis (tree), 187
liquid smoke, 157–59
Locard, Edmond, 200–201
Locard's Principle, 199–201
Lone Ranger, 195–96
Louis XIV (king of France),
    107
lovastatin, 112, 165–66
LSD, effect on spiders, 62
Luther, Martin, 79
lye, 181, 239–40

Mabury, Scott, 241, 242
MacDougall, Duncan, 208–10
magicians, 35–36, 50–51,
    93–95
magnesia alba, 18–19
magnesium, sound-absorbing
    properties, 193–94
magnesium carbonate, 18–19
magnesium hydroxide, 18–19
magnesium sulphate
    heptahydrate, 193–94
magnets "human," 92–95
Magola, Miroslaw, 93–95
Maguire Seven, 245–46
malaria, 73
maleic anhydride, 77
mangos, African, 147–49
Maple Leaf Gardens, 5
maps, printed on silk, 8–9
Marcone, Massimo, 137
marngrook, 204
Martin, Archer, 16–17
Masai people, 139
matches, 37, 40
mauve (synthetic dye), 24, 68
meat
    cured, 132–34
    "in vitro," 167–68
    kangaroo, 203–4
    "restructured," 161–64
"meat glue," 161–64
The Medical Journal of Australia, 14
medications
    anticoagulant, 170
    Aspirin, 121–22
    derived from moulds,
        109–11

injection of, 3–4
reaction with grapefruit juice,
    102
sourced from animal toxins,
    47–49
medicine shows, 96
Medina, Bartolomé de, 21–22
Merck (pharmaceutical
    company), 111–12
mercury
    and making of felt, 226
    as syphilis treatment, 223
    use in smelting, 21–22,
        227–28
Merlinite, 76
Meselson, Matthew, 57
"metaphysical jewellery," 75–76
methane, 203
Mevacor, 112, 165
mevastatin, 109–12
MI9 (British military intelli-
    gence), 8
Microsoft, 245
microwave ovens, 193
milk, sour, 140
"milk of amnesia," 123–25
milk of lime, 95–97
Mills, Heather, 74
Ming dynasty (China), 59
molecular gastronomy, 161–63
molecules, bonding, 253–55
Molotov, Vyacheslav, 232
Molotov cocktails, 232–33
monaculin K, 165
monamine oxidase, 53
monarch butterfly, 48
Monroe, Marilyn, 247

Moore, Demi, 73
Moreno y Maiz, Thomas, 4
morphine
    effect on spiders, 62
    injection of, 3–4
    side effects, 47–48
moulds, medications derived
    from, 109–11
mugariga plant, 87–89
Murray, Conrad, 123, 124
muscarinic acetylcholine
    receptors, 120
mussels, 86–87
Mwasapile, Ambilikile, 87–89
mycotoxins, 56–58
myelin, 117
myoglobin, 136
myrosinase, 168–69, 170

Nägeli, Karl Wilhelm von, 223
nanoparticles, 224–25
naphthalene, ix
National Aeronautics and Space
    Administration (NASA),
    62–63
National Center for Alternative
    and Complementary
    Medicine, 71–72
neoenergy, 81
Nero (Roman emperor), 23
neurotransmitters, 53, 68–69
Newton, Isaac, 199
Niemann, Albert, 4
ninhydrin, 33
nitrates and nitrites, 132–34
nitrocellulose, 16, 246
nitrogen, in Guinness beer, 28

nitroglycerin, 245–46
nitrosamines, 133
nitrous oxide, 197
Nixon, Richard, 166
Nobel Prize in Chemistry, 16,
    254
nuts, and weight control, 139

Obama, Barack, 50, 51
obesity, 147–49
Oliver, Jamie, 84–87
Olszak, Ryszard, 81
Olympic Games, 61–62
opiates
    injection of, 3–4
    side effects, 47–48
    use in patent medicines, 10
orangutans, 183
organ transplants, 103
organic chemistry, 254
ornithine, 104
Ostwald process, 206
ouabain, 48–49
ovotransferrin, 137
oxygen masks, in airplanes, 30
Oz, Mehmet, 149

painkillers
    derived from "poison dart"
        frogs, 47–48, 119–21
    injection of, 3–4
    morphine, 3–4, 47–48, 62
Palin, Sarah, 73–74
palm oil, 182–83
Palmolive soap, 181–83
paper chromatography, 16
para-dichlorobenzene, viii–ix

Pasteur, Louis, 143
patent medicines
    derived from petroleum, 201
    opiates in, 10
    Sequarine, 12–13
patio process, 22
"pearl ash," 140–41
pectin
    enzyme to break down, 110
    in printing inks, 8
Peking duck, 164, 166
Pemberton, John, 10
pencils, 256
Penicillum citrinum, 109, 111
pennies as means of defeating
    Breathalyzers, 113–14
pentane, 188
People for the Ethical Treatment
    of Animals (PETA), 51
pepper, black, ix
perfluorooctanoic acid (PFOA),
    241–42
perfluorooctanyl sulphonate
    (PFOS), 241
Perfume: The Story of a Murderer
    (Süskind), 186
Perkin, William Henry, 24, 68
Perricone, Nicholas, 176
Persil (detergent), 257–58
Peters, Hans, 63
petroleum, 201–2
    detergents derived from, 269
Pfizer (pharmaceutical
    company), 112
phenylethylamine, 52–53
pheromones, x–xi
phonograph records, 235–36

phosphates, 37
phosphorus, 35–37
photocatalysis, 31–32
photosynthesis, 145, 272–73
phthalates, 236–38
pigs
    pheromones, x–xi
    skin as model for human
      skin, 219–20
Pimat, 80–84
piperine, ix
placebo effect, 14, 83
planar sheets, 255
plaque, on teeth, 103–4
playing cards, 246
Plexiglas, 7, 247
"poison dart" frogs, 47–48,
    119–20
polar bears, 154
polycarbonates
    and bisphenol A, 213–15
    as scratch-resistant coating, 7
polycyclic aromatic hydrocar-
    bons (PAHs), 158–59
polyethylene, 187
polyfluorinated carboxylic acids
    (PCFAs), 242
polyfluoroalkyl phosphate
    esters, 240–42
polymer chemistry, 187
polymerization, 33–34, 78,
    247–48
polymethyl methacrylate, 7,
    246–49
polyolefin elastomer, 265–66
polypeptides, 27–28
polystyrene, 187–88

polyvinyl chloride (PVC),
236–38, 242–45
porcupines, 48–49
Post, Mark, 167
potassium bicarbonate, 234
potassium bromide, 150–51
potassium carbonate, 140–41
potassium chlorate, 29–30, 233
potassium cyanide, 62
potassium iodate, 106
potassium nitrate, 132, 205–6
potassium perchlorate, 198–99
potatoes
    french fries, 139
    green, 189
    processing into *chuno*, 22
Potts, Sir Percival, 158
"power prefixes," 259
Pravachol, 112
pravastatin, 112
Prince Rupert's Drops, 4–6
propofol, 123–25
Proposition 65 (California), 160
propylene glycol, 271
prostate cancer, 171
pseudo-science, 67–97
    endorsement by celebrities,
    71–74
Pure Food and Drugs Act
    (U.S.), 9–11
putrescine, 103–4
pyramids, purported power of,
    81–82

Quantum Xrroid Consciousness
    Interface Machine, 70
"quartz crystal singing bowl," 70

quercetin, 146
quicklime, 97

radiesthetic colours, 81
rain forests, 183
raisins, gin-soaked, 131–32
Randi, James, 93–95
rats, African crested, 48–49
red mud, 239–40
red yeast rice, 164–66
Reid, Alex, 72–73
Reid, Richard, ix–x
retinol, 154–55
rhabomyolysis, 166
rhinoceros horn, 54–56
Robiquet, Pierre-Jean, 108
rocket engines, 29–30, 199
rodents
    African crested rat, 48–49
    repellents, 52–53
Röhm, Otto, 247
Rome, ancient
    belief in aphrodisiacs, 107
    inks made in, 221
    use of amethyst, 74–75
    use of syringes, 3–4
rosacea, 176
Royal Armament Research
    and Development
    Establishment, 246
royal purple, 23–24
Royal Society, 6
rubber, vulcanized, 197
Ruhemann's purple, 33
Rupert (crown prince of
    Bavaria), 4–6
rutin, 146

Sade, Marquis de, 107–8
St. John's wort, 103
saliva
    microbes in, 103
    pheromones in pig's, x
    of swiftlets, 135
    testosterone in, 206–8
salt
    Epsom (see Epsom salt)
    in fingerprints, 33
    iodized, 106
    mining of, 201
saltpetre, 132, 205–6
Sankyo (pharmaceutical
        company), 110, 111, 112
Sapindus mukorossi (tree), 188
saponins, 146, 189–90, 269
"Satan's telegram" (magician's
        illusion), 35–36
Sawalha, Julia, 73
Scheele, Carl Wilhelm, 62
schizophrenia, 63–64
Schönbein, Christian Friedrich,
        16–18
Schwetzingen (Germany), 145
Scientology, 73
sebum, 184
self-cleaning windows, 30–32
Sequarine, 12–14
Seveso (Italy), 243
shampoo
    baby, 178–80
    two-in-one, 184–85
Sharpe, Richard, 213, 215, 216,
        218–19
shellac, as food additive, 85
silk, printing on, 8–9

silver
    amalgam, 228–29
    antimicrobial properties,
        223–24
    colloidal, 225–26
    smelting of, 20–23
    tarnishing of, 195
silver nitrate, 33, 225
Simon, Eduard, 187
simvastatin, 112
Singh, Sawai Madho II, 223
skatole, 41
skin cancer, 158, 176–77,
        248–49
skincare products, 175–78
"skunk bombs," 41
Skylab, 62–63
Smalley, Richard, 254
smelting
    of aluminum, 239–40
    of gold, 227–28
    of silver, 20–23
Smith, William, 202
smoke
    from burning wolf dung,
        58–60
    from indoor fires, 60
    liquid, 157–59
soap, 181–83, 257, 269
soap nuts, 188–90
soapwort, 269
Sobel, Noam, 206–8
sodium bicarbonate, 122, 141,
        234
    as salt additive, 106
sodium hydroxide, 181, 239–40
sodium perborate, 258

sodium silicate, 258
sodium thiosulphate, 106
soft drinks, 9–11, 140, 149–51,
     152, 159–61
solanine, 189
soluble fibre, 148
Somers, Suzanne, 71–74
soot, 268
soul, existence of, 208–10
sound waves, absorption by
     magnesium, 193–94
soursop, 89–92
Soviet Union, 56–58
space shuttle, 199
Spanish fly, 107–9
sparklers, 198–99
"spice effect," 170–71
spiders, experiments on, 62–64
Stachybotrys chartarum, 57
stanol esters, 143–44
statins, 109–13
stem cells, ix, 155, 167
sterol esters, 143–44
stink bombs, 39–41
storax, 186–87
Stork, Gilbert, 108
Straub tail response, 120
Strength Within (anti-wrinkle
     supplement), 175–78
Streptoverticillium mobaraense, 162
stress, "geopathic," 82
Stuart, George, 11
styrene, 187–88
Styrofoam, 188
styrol, 187
sublimation, 33
succinic acid, 76–80

sulphoraphane, 168–71
sulphur dioxide, 195, 267–68
sulphuric acid, 231, 233, 267,
     268
superhydrophilicity, 31
Suppes, Galen, 270–71
surface tension, 189
surface-active compounds,
     27–28
surfactants, 184
     "green," 269–70
Süskind, Patrick, 186
sweat
     androstenone in, x–xi
     and cutaneous adhesion
          syndrome, 94
     and fingerprints, 33
     odour of, 104
swiftlets, 135–37
Sylvius, Franciscus, 130–31
Synge, Richard, 16–17
synthetic dyes, 24
syphilis, 223, 225
syringes, hypodermic, 3–4

Tainter, Charles Sumner, 236
Tang (orange drink crystals),
     229–31
tannic acid, 221–22
Teflon, 241
tempered glass, 4–7
Tenkaev, Leonid, 92–93
terpenes, 78, 269
terpinen-4-ol, 131
testosterone, 14, 206–8
tetrafluoroethane, 265–67
thermal printing paper, 216,

219–21
thin-layer chromatography,
    245–46
3M (manufacturing company),
    241
Thurston, Howard, 50–51
Thurston, Rae, 50–51
thyroid gland, 105
Times Beach (Missouri), 243
titanium dioxide, 30–32
Titusville (Pennsylvania), 201
Tofana, Giulia, 108
Tom Jones (Fielding), 130
Toxicological Science, 218
toxins, elimination through
    kidneys, 102–3
traditional Chinese medicine,
    54–56, 136, 164–66
transglutaminase, 161–64
tree frogs, 47–48, 119–20
triacetone triperoxide (TATP),
    229–32
trichothecenes, 56–58
trophy hunting, 55
truffles, x–xi
Tsvet, Mikhail, 15, 18
tumours
    and creosote, 158
    and ferulic acid, 146
    and formaldehyde, 179
    formation, ix
    injection with titanium
        dioxide, 32
    prostate, 171
21 Grams, 210
Tyrian purple, 23–24

ultraviolet light, 248
Unilever, 175
United States
    Bureau of Chemistry, 9–10
    Department of Energy,
        215–16
    Food and Drug
        Administration, 46,
        165–66, 215–16, 217
    Pure Food and Drugs Act,
        9–11
    space program, 38, 62–63,
        199, 224
    U.S. Government Standard
        bathroom malodor, 41
*United States vs. Forty Barrels and
    Twenty Kegs of Coca-Cola*,
    9–11
unnatural wavelength effect
    (UWE), 68–69
urine
    phosphorus in, 36–37
    as rodent repellent, 52–53

vegetable oils
    brominated, 149–51
    in soaps, 181–83
Vensal, Richard, 145–46
Viagra, 54–56
Victoria (queen of England), 24
vinegar, 96–97
vitamin A, 154–55
vitamin B$_{12}$, 116–19
vitamin C, 176
vitamin D, 19
vitamin E, 176
vitamin K, 170

volcanoes, 195, 243, 257, 267
Volta, Alessandro, 61, 228
voltaic piles, 61, 228
vom Saal, Fred, 214–15
von Hippel, Bill, 54
von Hippel, Frank, 54
vulcanized rubber, 197

Waddingtons. *See* John
    Waddington Ltd.
warts, treatment of, 107–9
water, purification of, 223–25
Watson, Jonathan, 202
WD-50 (restaurant), 161, 162
Weizmann Institute of Science,
    206–8
Who Me? (foul-smelling spray),
    41
widgets, in canned beer, 27–29
Wiley, Harvey Washington,
    9–10
William III (king of England),
    129–30
windows, self-cleaning, 30–32
wireless Internet, purported
    dangers, 70

Wischmeyer, Paul, 124
Witt, Peter, 63
Wöhler, Friedrich, 254
wolf dung, 58–60
Wood, Alexander, 3–4
Woodward, Robert, 116
World War II
    espionage, 34–35
    phosphorus bombs, 36
    printing of maps on silk, 8–9
    Who Me?, 41
Wright, John, 62

xylan, 157
xylitol, 156–57

Yamagiwa, Katsusaburo, 158
"yellow rain," 56–58
yenolab receptors, 69
yogurt, 139–40, 144

Zalko, Daniel, 219
zeolites, 264
Zhou Enlai, 166
Zocor, 112